Antique & Collectible
Dollhouses
and Their Furnishings

Dian Zillner & Patty Cooper

4880 Lower Valley Road, Atglen, PA 19310 USA

This book is dedicated to Paige Thornton, who learned about dollhouse furniture the hard way, one piece at a time, then so generously shared that knowledge with all of us over the past twenty-five years. She is representative of all the dollhouse collectors and dealers who so willingly help put together pieces of the puzzle in order to answer the ongoing questions of "Who made that and when?"

Copyright © 1998 by Dian Zillner and Patty Cooper
Library of Congress Catalog Card Number: 97-80141

All rights reserved. No part of this work may be reproduced or used in any form or by any means—graphic, electronic, or mechanical, including photocopying or information storage and retrieval systems—without written permission from the copyright holder.

Designed by "Sue"

ISBN: 0-7643-0120-9
Printed in China
1 2 3 4

Published by Schiffer Publishing Ltd.
4880 Lower Valley Road
Atglen, PA 19310
Phone: (610) 593-1777; Fax: (610) 593-2002
e-mail: schifferbk@aol.com
Please write for a free catalog.
This book may be purchased from the publisher.
Please include $3.95 for shipping.

Please try your bookstore first.

We are interested in hearing from authors
with book ideas on related subjects.

Notice: All of the items in this book are from private collections and museums. Grateful acknowledgment is made to the original producers of the materials photographed. The copyright has been identified for each item whenever possible. If any omission or incorrect information is found, please notify the authors or publisher and it will be amended in any future edition of the book.

The prices listed in these captions are intended only as a guide, and should not be used to set prices for dollhouses and related products. Prices vary from one section of the country to another and also from dealer to dealer. The prices listed here are the best estimates the authors can give at the time of publication, but prices in the field can change quickly. Neither the authors nor the publisher assume responsibility for any losses that might be incurred as a result of consulting this price guide.

TABLE OF CONTENTS

Introduction 4	German Wood Dollhouse Furniture 106
American Lithographed Dollhouses and Furniture 6	German Metal Dollhouse Furniture 120
R. Bliss Manufacturing Co. 6	**Other Wood Dollhouses and Furniture** 124
A Series of Lithographed Houses by an Unknown Maker 23	Hall's Lifetime Toys 124
Whitney S. Reed Company 27	Lynnfield-Block House-Sonia Messer 129
Dunham's Cocoanut 29	Menasha Woodenware Corporation 135
McLoughlin Brothers 31	Mystery Houses and Other Houses Sold by F.A.O. Schwarz 137
F. Cairo 34	Wisconsin Toy Company 140
Grimm & Leeds 35	Miscellaneous Wood Dollhouses (And Some Furniture) 145
English Dollhouses and Furniture 36	Miscellaneous Wood Dollhouse Furniture 163
G. & J. Lines and Lines Brothers 36	**Metal Furniture** 185
Box Back Dollhouses Often Known as Silber and Fleming Types 48	**Paper and Cardboard Dollhouses and Furniture** 198
Amersham Works, Ltd. 50	**Plastic Furniture and Metal Houses** 217
Hobbies 52	Reliable 217
Tudor Toy Company and Others 54	Miscellaneous Furniture and Houses 221
A. Barton and Co. (Toys) Ltd. 57	**Miscellaneous** 237
Dol-Toi Products (Stamford) Ltd. 67	Accessories 237
Other British Dollhouse Furniture 71	Dollhouse Dolls 251
German Dollhouses and Furniture 76	**Sources** 266
Moritz Gottschalk 76	**Bibliography** 268
Christian Hacker 100	**Index** 270
D.H. Wagner & Sohn and German Dollhouses of Unknown Origin 103	

INTRODUCTION

The authors, Dian Zillner and Patty Cooper, became acquainted during the writing of *American Dollhouses and Furniture From the 20th Century*. After that book's publication, the two collectors continued to find information on commercially produced dollhouses and their furnishings. This book is the result. It is meant to be used as a sequel to the first volume and features older dollhouses and furnishings from the United States as well as items from Germany, England, and Canada. It also includes information on American companies not covered in the previous book and some updates on companies discussed earlier.

Although commercial dollhouses and rooms were produced throughout most of the 19th century, it was not until the late-1800s that techniques of mass production allowed dollhouses to become a viable part of the toy industry. Germany led the world, both in the late 19th century and the early 20th century, in the exportation of dollhouses and furniture. Firms in England also became more active in the production of dollhouses and furniture during the mid to late-1800s. Perhaps inspired by their European counterparts, American toy companies such as Bliss, McLoughlin, and W.S. Reed began adapting the process of applying lithographed papers over wood as an economical way to create dollhouses that could be sold to the growing middle class.

It is hoped that this book will continue to expand knowledge about the companies which have produced dollhouse related toys from the 19th century through the 1970s. With over 800 color photographs, this book, along with the earlier volume, should enable collectors to identify the products of many different manufacturers, including several that were previously unknown.

As with all collectibles, the value of a dollhouse or furniture depends on the condition, popularity, and rarity of the piece. A collector will pay more for a mint-in-box set of furniture than for the same furniture assembled one piece at a time. Dollhouses in original condition, with no repainting, replaced parts, or replaced papers, will command the highest prices. Missing doors, windows, and chimneys substantially lower the price of a house. Damage to a house or insensitive restoration also decreases its value.

The many pictures in this book are intended to illustrate the wide variety of items which are available to today's collectors. Included in this publication is a list of addresses to help collectors in their search for dollhouses and furniture. There is also a bibliography of materials used to research this volume which will provide helpful sources for those collectors who want to continue to explore the field.

Because of the assistance given by over fifty collectors, this book will provide much needed new information on dollhouses and dollhouse furniture. The authors gratefully acknowledge the many individuals and organizations who answered questions, shared materials, and took photographs in order to make this book possible: Vivia Allor, Judith Armitstead, Sharon and Kenny Bernard, Linda Boltrek, John Blauer, Mary Brett, Gail and Ray Carey, Annette Clark, Shirley Cox, Bill Dohm, Karen Evans, Zelma Fink, Nick Forder, Lois L. Freeman, Kathy Garner, Kibbe Gerstein, Rita Goranson, Rosemary Grant, Linda Hanlon, Mary Harris, E. Munroe Hjerstedt, Sally Hofelt, Ann Hurley, Flora Gill Jacobs, Jeanne Kelley, Lisa Kerr, Madeline Large, Left Bank Antiques, Jenny LeMasurier, Louise McCauley, Carol Miller, Joan and Gaston Majeune, Renee Majeune, Margaret Woodbury Strong Museum library, Bob Milne, *Miniature Collector Magazine*, Nanci Moore, Marge Meisinger, Arliss Morris, Judith Mosholder, George Mundorf, N.D. Cass Co., Neenah Historical Society, Betty Nichols, Becky and Don Norris, Ann Parker, Leslie and Joanne Payne, Ruth Petros, Marilyn Pittman, Elaine Price, Jill H. Ramsey, Geraldine Raymond-Scott, Nancy Roeder, Joanie Searles, Roy Sheckells, Marion Schmuhl, Julie Scott, Suzanne Silverthorn, Louana Singleton, Roy Specht, Linda and Carl Thomas, Anne B. Timpson, Paige Thornton, Steve Thornton, The Toy and Miniature Museum of Kansas City, Marcie and Bob Tubbs, Mary Lu and Bob Trowbridge, Sharon Unger, and Mildred Felder Whitmire.

Acknowledgement and recognition is also extended to Schiffer Publishing Ltd. and its excellent staff, particularly to Sue Taylor, layout editor, and editor Dawn Stoltzfus who helped with this publication. Without their support and extra effort, this book would not have been possible.

The following people are among those who so generously helped in the preparation of this book:

Gail Carey, who so generously shared her collection in order to help make this book possible, is shown with the authors, Patty Cooper and Dian Zillner, at the 1996 *Dollhouse and Miniature Collectors Quarterly* conference.

Flora Gill Jacobs is responsible for the core body of knowledge in the history of American dollhouses. Through her many books, articles, and the establishment of the Washington Dolls' House and Toy Museum, she has shared her extensive knowledge of the field and been an inspiration to dollhouse collectors around the world.

Collector Linda Hanlon contributed many photos of wood furniture and dolls from her collection. Mary Brett, author of *Tomart's Price Guide to Tin Litho Dollhouses and Plastic Dollhouse Furniture*, generously shared her research. Lois Freeman knows more about fiberboard houses than almost anyone and also shared her extensive collection of 1" scale furniture and Gottschalk houses.

Roy Specht provided many of the unusual plastic pieces for this book. George Mundorf, with his love of Art Deco, shared many of his finds. Kathy Garner hosted the first *Dollhouse and Miniature Collectors Quarterly* conference which allowed collectors from all over the country to meet and share their enthusiasm. Sharon Unger, editor of *DMCQ*, continues to provide a forum for exchanging information on the hobby.

Marcie Tubbs has written several articles on plastic dollhouse furniture and provided many photographs for this book. Judith Mosholder, a collector par excellence of plastic dollhouse furniture, provided much needed information. Marilyn Pittman always seems to find the mystery pieces which continue to make dollhouse collecting fun.

Becky Norris shared her extensive collection of cardboard houses and advertising items. Ruth Petros managed to find obscure examples to help fill in some of the blanks for this book. Patricia Flenker attended the DMCQ conference and shared information with everyone. Nanci Moore provided many examples from her collection and offered her expertise on plastic furniture and cardboard houses.

AMERICAN LITHOGRAPHED DOLLHOUSES AND FURNITURE

At the end of the 19th century American toy companies began to show a greater interest in manufacturing dollhouses. Before this time, most of the known dollhouses were handmade and available only to children of the upper class. Following the example of German companies such as Gottschalk, several American firms began to use the technique of applying lithographed paper over a simple wood structure to create dollhouses which offered a richness of architectural detail but were economical to produce.

R. BLISS MANUFACTURING CO.

The R. Bliss Manufacturing Co. was founded by Rufus Bliss in 1832 to make wood screws and clamps. The company made many different products over the years and by 1871 they were making toys. Although Rufus Bliss died in 1879, the company continued using his name and went on to become a leader in the production of toys, especially those made of lithographed paper over wood. The Bliss company entered the dollhouse market gradually. Their first known dollhouse was "The Fairy Doll's House" advertised in 1889. This was a simple, open-fronted structure formed by two boxes, the smaller of which could be stored inside the other. Like most other Bliss toys, the house was covered in elaborately lithographed paper. An open-backed, two-story house with a bow window was introduced in 1895. It sold for $1.00 furnished with ten pieces of furniture.

A few other dollhouses were added during the 1890s, but the houses from the 1901 catalog are the ones most readily identified as Bliss. The seven houses in this series were mostly Queen Anne in style with elaborately printed "gingerbread" trim, turned posts, balconies, wrap-around porches, and an occasional turret. Many of the windows were cut out and had mica or isinglass panes. They were marked with the name "R. Bliss" inside an elongated "C" with an "o" at the end (for "company"), a logo which vaguely resembled a fish. In obscure places along the edges of the exterior papers, a discerning eye can usually find a three-digit number which corresponds to the number of the house in the catalog. The numbers in the 1901 series ranged from 570, the smallest, only 9.5" high, to 576, which was 28.5" high at the top of its chimney. The smaller houses were produced in great quantities and shipped to retailers in crates containing two to six dozen houses. The "Tower House," number 576 in the series, was the largest and most elaborate, with a wraparound porch, three gables, five turned posts, and a six-sided turret topped with a turned finial. It was also the most expensive to produce, with a wholesale price of $5.00, a dollar more than the next closest size. These were shipped to retailers with only six in a crate. It would be safe to assume that few of the largest houses were made, and consequently, the "Tower House" is rare today and very expensive. One was sold at auction in 1990 for $11,500, approximately three or four times the expected price for any other Bliss house at that time. Many of the houses from the 500 series were produced throughout the life of the company, concurrent with newer models.

Thus far, no Bliss catalogs have been found for the decade between 1901 and 1911, but a few houses have survived which appear to be from that period. These include two small cottages, number 613, the "Bird House," and number 616, the "Garden House," as well as the "Adirondack Cabin" in two- and four-room versions. The A.C. McClurg & Company's catalog for 1909-10 shows the two-room version of the Adirondack Cabin, complete with deer head gable ornament, along with three other Bliss

This house was shown in the 1895 Bliss catalog as No. 240 and was said to retail for $1.00. It came with ten pieces of furniture which appear to be the same as those marked 217 (shown later in this chapter). It is open-backed and contains two rooms with richly lithographed interiors which show draped windows, flowered wallpapers, portraits of children, and a roaring fire. The front of the house features a rounded "bow window," which gives the house a nickname by which it may be identified by collectors ($2000-3000). 21" high x 11" wide x 13" deep. *House from the collection of Carl and Linda Thomas. Photograph by Patty Cooper.*

houses. This provides support for the convincing case made by Flora Gill Jacobs in *Dolls' Houses in America* that the Adirondack Cabin, although unmarked and unnumbered, was made by Bliss. It should also be noted that some examples of the Adirondack Cabin contain floorpapers and wallpapers that have been found on other, marked Bliss dollhouses.

A Butler Brothers advertisement, said to be from 1910, was reprinted in *Dolls' Houses in America* showing four other unmarked Bliss houses. These houses are something of a mystery in that they have metal railings, usually found on later houses, rather than the lithographed balustrades used on Bliss houses at least through 1911. Also, examination of the actual houses reveals that at least two of them bear numbers which were previously used in the 1901 catalog for completely different designs. These numbers are a bane to the complacent researcher, upsetting what would otherwise be a logical sequence which could be used to date Bliss houses and link them to specific catalogs.

The Bliss catalog for 1911 was reprinted by Blair Whitton as part of the book *Bliss Toys and Dollhouses.* This catalog showed seven different dollhouse designs, along with one which was advertised in the 1910 Butler Brothers catalog. Four of the new dollhouses were also available in folding versions. The dollhouses in this series ranged from number 200, a two-room model only 10" high, to number 207, which was 24" high and contained four rooms. These houses were simpler in construction than earlier houses. Many of the windows were lithographed rather than being cut-out and lithographed paper was used to cover straight-sided posts in place of the turned pillars. None of the houses shown in this catalog were marked with the full Bliss name, although number 206 had a logo which contained a "B."

Thus far no Bliss catalog dated later than 1911 has been found. In 1914, the company was sold to Mason and Parker who continued to manufacture many of the same dollhouses as late as the 1920s.

No shortcuts, such as using the same window design or porch rail for several houses, were taken in designing the lithography for any of the houses shown in the known Bliss catalogs. This is in contrast to houses made by Converse or the unknown "Gutter House" company in which the same elements were repeated on various models. However, the papers designed for the early houses have been found, in various combinations, on other, differently constructed, dollhouses. These include several four-room houses, some with two doors which give the appearance of duplexes. The provenance of these delightful, but uncataloged houses remains a mystery. The most logical theory is that these houses were actually produced after Bliss was sold to Mason and Parker in 1914. Perhaps the new buyers inherited the factory, parts, lithographic plates, and remaining stocks of printed paper, but not the artists. It would have been an effective marketing strategy to simply apply the existing papers to newly designed boxes which often contained more rooms but were more streamlined in shape. In this way, the new houses would have had the greater appeal of added play space, appeared more contemporary in style, and been easier to construct, all without the expense of paying artists or producing original plates. Although quite attractive, these later versions are clearly hybrid houses, incorporating parts from various models in the 500 and 600 series. The paper doesn't always fit quite right and was obviously pieced. Sometimes the placement of the doors and windows doesn't really make sense and the symmetry is a little off. These houses almost invariably have lithographed, rather than turned, posts and pierced metal, rather than lithographed, balustrades.

Although unmarked and not shown in any of the known catalogs, this house has all the characteristics of a Bliss. It has bright, four-color lithography with the number 342 printed on the edge of its papers. The porch posts appear to be the same as those on the marked Bliss Wild Rose Cottage and similar to those on Bliss No. 571. The front of the house opens in one piece on the kind of butterfly-shaped hinges used by Bliss ($1400-1600). 17.5" high x 11" wide x 8.75" deep. *House from the collection of Shirley Cox. Photograph by Sarah Cox.*

The interior of Bliss No. 342 is very similar to Bliss No. 240 from the 1895 catalog. It has richly lithographed details which include wallpapers, draped windows, framed pictures, and a roaring fire. The presence of turned columns, printed architectural features on the interior walls (as opposed to wallpaper), and a number in the 300s indicate that this house is circa 1896-1900. *House from the collection of Shirley Cox. Photograph by Sarah Cox.*

8 American Lithographed Dollhouses and Furniture

It is somewhat frustrating to try to decide if an unmarked house might be Bliss. Rufus Bliss died long before the first Bliss dollhouse was ever manufactured and there is no way of knowing how many different artists might have had a hand in designing the dollhouses over the years. Therefore, some variations in style are to be expected, and do, in fact, seem to be detectable. Bliss houses can be characterized by the high quality of the lithography, employing the full four-color, spectrum of colors in an almost whimsical way. The interiors of the houses rarely live up to the extravagant decoration of the outside. The smallest contain only one or two rooms and the largest known Bliss house only has four. The interiors of early houses appear to be papered in "real", full-size wallpapers, but later houses, from at least 1911 onward, have smaller scale prints which were probably designed specifically for the dollhouses. There are a few other clues to look for. As mentioned earlier, even the unmarked Bliss houses usually have an obscure three-digit number printed on the edges of the papers. Most of the hinges, both those that are attached to doors and those which allow the houses to open, are shaped somewhat like a butterfly, with a little extra point extending between each lobe. This shape is distinctive when compared to the hinges used by Gottschalk or the unknown "Gutter House" company, but unfortunately, indistinguishable from those used by Whitney S. Reed or Converse.

The smallest house from the 1901 catalog, No. 570 has also been found with a projecting, wood roof over the porch and opening door. The small print wallpapers and lack of opening door suggest that this example was made later. The house contains two rooms with ceilings only 3.5" high ($700-800). 9.5" high x 7" wide x 4" deep. *House and photograph from the collection of Patty Cooper.*

The name "Wild Rose Cottage" is printed over the door of a house that is marked Bliss on one side, along with the number 395. The front pictures a girl knocking at the door, a whimsical feature not seen on other Bliss houses. The size of the girl in relation to the rest of the house indicates that perhaps Wild Rose Cottage was intended to represent a child's playhouse rather than a residence. It is believed that this house, with its similar porch posts and related number, was probably introduced the same year as Bliss No. 342. However, Bliss began another numerical sequence around 1910. It is possible that this open-backed house, which is papered in a small print, could be from a later time. Until more catalogs are found, this will remain a mystery ($1000-1200). *House and photograph from the collection of Carl and Linda Thomas.*

Bliss No. 571 was listed in the 1901 wholesale catalog for 50 cents. This example has an opening door and four cut-out windows. At least two variations have been found. One has a porch which extends all the way across the front of the house and only two cut-out windows. Another, later version, has lithographed, rather than turned posts, no cut-out windows, and a lithographed pediment over the entryway ($800-1200). 12.5" high x 9" wide x 6.5" deep. *House and photograph from the collection of Patty Cooper.*

American Lithographed Dollhouses and Furniture 9

No. 572, known to collectors as the "Keyhole House," was listed in the 1901 catalog as the $1.00 size. It had a significantly more complex construction with projecting dormers on each side of the roof, latticed gable with cut-out keyhole decoration, and elaborately turned posts. Inside, there are two rooms with ceilings approximately 6" high. This is the only Bliss house known to have come in more than one color combination. The one shown here has soft yellow "clapboard siding" with pink and turquoise trim. Another version has been seen in which the body of the house is printed in a rose color rather than yellow. The rose-colored version is usually found with a box base covered with lithographed paper rather than the plain, open-fronted base of the model shown. The chimney is missing on this house ($1200-1600). 16" high x 10" wide x 7" deep. *House and photograph from the collection of Patty Cooper.*

The interior of No. 571 is papered with an overscale pattern which appears to be "real" wallpaper, an indication that this is one of the earlier houses. There are two rooms, with ceilings approximately 4.5" high. *House and photograph from the collection of Patty Cooper.*

The "Bay Window House" was shown as No. 573 in the 1901 catalog, at a wholesale price of $2.00. The exterior is printed with orange brick on the first floor and white clapboard siding on the second. Like the Keyhole House, it has a shed dormer on each side of the steeply pitched roof. Although it only has two rooms, this model, with ceilings 7" high, is large enough to furnish with small-scale Biedermeier furniture ($1600-2000). 18" high x 10.5" wide x 8.5" deep. *House and photograph from the collection of Patty Cooper.*

This "Large Cottage" which makes use of the orange brick and white clapboard paper of No. 273, the Bay Window House, was advertised in a circa 1910 Butler Brothers catalog reprinted by Flora Gill Jacobs in *Dolls' Houses in America*. On this variation, two of the windows from the original bay have been flattened out and placed on their sides above the second story balcony. The paper which originally adorned the side dormers is now found on two hipped-roof dormers at the front of the house. This house is slightly wider than its predecessor and the paper on the base shows the piecing that was necessary to achieve this ($1000-1400). *House and photograph from the collection of Patty Cooper.*

Bliss No. 574 was described in the 1901 catalog as a "beautiful seaside residence," therefore collectors usually refer to it as the "Seaside." For a wholesale price of $3.00, it incorporated many of the features found on the smaller houses from the same year. Like the Keyhole house, it has a lattice panel in the front gable, but on this model the cut-out is squared. There are three gable-roofed dormers, a second story balcony, a porch which extends across the front of the house, and five turned posts. The exterior of the first floor is printed in a yellow clapboard and the second story has pink shingles. It is marked with the Bliss name on both doors and on the pediment of the main entrance ($2500-4000). 20.5" high (to top of roof) x 18" wide x 10" deep. *House and photograph from the collection of Patty Cooper.*

Perhaps the most intriguing attribute of the Seaside is that it contains three rooms, an impressive number for an early Bliss. The house opens from the front to provide access to the first floor parlor and upstairs bedroom. It also has an opening on the left side into what was clearly meant to be the kitchen, complete with service entrance. The parlor is papered with the kind of overscale, "real" wallpaper typical of the early Bliss houses. A similar paper is used on the floors. The upstairs paper may be a later replacement. *House and photograph from the collection of Patty Cooper.*

American Lithographed Dollhouses and Furniture 11

This four-room house, with a center gable, is a later variation of Bliss No. 574, probably made after 1914. The yellow clapboard and pink shingled papers, originally designed for the Seaside, have been pieced together to cover a larger box, which is much simpler in construction. Although the house is a symmetrical design, the paper doesn't quite fit that way. In order to cover the front, the printed windows which were on the sides of the Seaside, were placed on the front of "Center Gabled House" and the door is off-center. The turned columns have been replaced with more economical strips of wood covered with lithographed paper. The balcony rails are made of a combination of pierced metal (common on these later houses) and lithography. The lithographed portion on the second story incorporates the balustrade design from Bliss No. 576, known to collectors as the "Tower House." It is embellished by a decorative cornice which originally was used on the gate of Bliss No. 616, the "Garden House." The house opens on both sides ($2000-3000). 21" high x 20" wide x 11" deep. *House and photograph from the collection of Patty Cooper.*

Described as "an elegant suburban home" in the 1901 catalog, Bliss No. 575 sold for $4.00, but contained only two rooms. The exterior of the "Suburban Home" has lithographed cut stonework in shades of gray and brown on the first story and orange brick on the second story. There are three dormers projecting from the front of the steeply pitched, lithographed roof and a porch which wraps around three sides of the house ($2500-4000). 24" high (to top of roof) x 18.5" wide x 11.5" deep. *House and photograph from the collection of Patty Cooper.*

Another variation of No. 574, circa 1914 or later, could be described as a "Semi-detached" pair. Because this model makes no attempt at symmetry, the Seaside paper is a better fit, except for the fact that the second story windows are partially obscured by the porch roof. The paper originally designed for the side dormers of the Seaside is used on the front gable of this house. Pierced metal is used for the railings and the porch columns are covered with lithographed paper instead of being turned. Each side of the house opens to provide access to four, connected rooms each papered in a small-scale print ($2000-3000). 20" high (to top of roof) x 19.5" wide x 11.5" deep. *House and photograph from the collection of Patty Cooper.*

"Square Towers" is a circa 1914 or later variation of Bliss No. 575. Its rather complex construction makes it an exception to the rule that these later houses were designed to be more economical to produce. Like other houses in which the 1901 papers were applied to different boxes, Square Towers has a metal balcony railing and square posts (although the lithography is missing from these). However, the two towers, with their turned finials, and the dormer on the roof provide a lot of architectural interest. The paper used for the dormer windows was originally designed for the sides of Bliss No. 616, the Garden House and the "ball-and-stick" pattern of the first floor balustrade is from No. 576, the Tower House. The house opens from each side and contains four rooms with connecting doors ($2500-4000). 23" high (to top of roof) x 21" wide x 12" deep. *House and photograph from the collection of Patty Cooper.*

No example is shown of Bliss No. 576, which was described in the 1901 catalog as "a modern city residence," but is usually known to collectors as the "Tower House." The Tower House was very complex in construction with five turned posts, three gables, a wrap-around porch, and a six-sided turret topped by a turned finial. This would have been a very expensive house to build as is evidenced by the wholesale catalog price of $5.00. It is therefore not surprising that very few examples of this house are still in existence. Even the later models which used the same paper are rare. The circa 1914 or later variation shown here takes the form of a duplex with a center gable. Once again, the placement of the paper does not quite match the symmetrical design of the house. Windows which were intended for the side of the tower house appear a little too low when applied to the front of the Duplex. The arched lattice work in the front gable is the same paper originally designed for the Keyhole House ($4000-6000). 25.5" high (to top of roof) x 20.5" wide x 14" deep. *House and photograph from the collection of Patty Cooper.*

In addition to being rare, this variation of No. 576 is intriguing because it is one of only two known Bliss models with a stairway. The house opens from each end and contains four rooms. The two upstairs rooms are connected by an opening door which is covered with the same paper used for exterior doors. A draped, keyhole shaped opening connects the two first floor rooms. The interior is papered with the kind of small-scale prints usually found in these later houses. The furniture shown is Bliss in approximately 1" to the foot scaie. *House and photograph from the collection of Patty Cooper.*

Bliss No. 613 is known as the "Bird House" because of the tower-like projection on the right side of the roof which closer examination reveals to be a birdhouse or dovecote. Although not shown in any of the known catalogs, this small house is marked "R. Bliss" inside an elongated "Co" (for company) design, which vaguely resembles a fish. This is the same emblem used on the doors of all of the dollhouses from the 1901 catalog. The use of the same logo and the proximity of the numbers suggest that this house may have been made circa 1902-1905 ($1000-1300). 11.5" high x 11.5" wide x 6" deep. *House and photograph from the collection of Patty Cooper.*

Another Bliss, which is similar in size, has been nicknamed the "Garden House." The number 616, printed on the edges of its lithographed papers, invites longing speculation as to the existence of dollhouses numbered 614 and 615. This two room house is obviously a contemporary of the "Bird House" and bears the same fish-like Bliss logo on the fascia above the porch. Unlike many Bliss houses, this one is open-backed. The house is separate from the base which contains the walled garden. Two small dowels are attached to the base and can be fitted through holes in the lower floor of the house to stabilize it ($1200-1500). The dimensions of the house, without the base, are 14" high (to top of roof) x 11" wide x 7" deep. The base is 15.5" wide x 12.5" deep. *House and photograph from the collection of Patty Cooper.*

One other open-backed house, which has been identified as Bliss, is the one known as the "Adirondack Cabin." Flora Gill Jacobs makes a strong case for this being a Bliss house in her book *Dolls' Houses in America*. This theory is further reinforced by the fact that a two-room version, with the same deer head adorning the front gable, is advertised in an A.C. McClurg & Co. catalog from 1909/10 along with three other known Bliss dollhouses. Floor papers and wallpapers from this house have also been found on other marked Bliss houses. However, it is the only house, reliably attributed to Bliss, which, when examined inch-by-inch with a magnifying glass, failed to reveal an obscure three-digit number. This house contains four rooms, one of which is accessible by lifting the American Indian ornament on the roof of the dormer ($2500-3000). *House and photograph from the collection of Patty Cooper.*

The "Colonial Mansion" is shown in a circa 1910 advertisement from Butler Brothers reprinted in Flora Gill Jacobs' *Dollhouses in America*. The double front doors lead into an arched opening which provides the dolls' access into two very tall, narrow rooms. Each of the sides opens to provide access for humans. Lest it be assumed that all Bliss numbers proceed in a logical and decipherable sequence, it should be noted that the gable paper on this house contains the number 573 1/2 in mirror image. The number 573 is the one found in the 1901 catalog for the Bay Window House although the lithography of the two houses is completely different. The Large Cottage variation of the Bay Window House, which also bears the number 573, is shown in the same Butler Brothers advertisement from 1910 ($1600-2000). 18.5" high x 16.5" wide x 10" deep. *House and photograph from the collection of Roy Sheckells.*

Bliss No. 200 is shown in the 1911 catalog reprinted as part of the book *Bliss Toys and Dollhouses* by Blair Whitton. The dollhouses in this catalog are much simpler in construction and would have been more economical to produce than earlier ones. All of the architectural detailing on this small house is provided by the lithography, including the trompe l'oeil bay window. The flat facade has no turned pieces or cut-out windows ($600-750). 10" high x 7.75" wide x 4" deep. *House and photograph from the collection of Patty Cooper.*

Although not shown in any of the available catalogs, "The Golden Rod" has all the characteristics of a Bliss including the obscure number 153 on the edge of the lithographed papers. The front opens on a cloth hinge and there are two interior rooms. Its number would suggest that it predates the houses in the 1911 catalog by a year or two. The fact that this house was given a name is interesting and may reflect, as dollhouses often do, trends in the "real world." During the early 1900s, several companies were marketing full-size, mail order houses. Active among them was Sears, who gave appealing names to many of the houses shown in their catalogs, including a one-story cottage, advertised in 1911, named "The Goldenrod." ($850-1000). 12" high x 18" wide x 9" deep. *House and photograph from the collection of Patty Cooper.*

American Lithographed Dollhouses and Furniture 15

The inside of No. 200 is papered with an oversize illustration of a draped window, which illogically extends through both the first and second story rooms. This paper was originally designed for a much larger house and can be seen in the "Unmarked Bliss" shown in *Dolls' Houses in America*. The side walls are papered in a small print, typical of later Bliss houses. The inside of the opening front shows the faux wood-grained paper often used by Bliss on the backs of dollhouses and as wallpaper in the Adirondack cabins. *House and photograph from the collection of Patty Cooper.*

The slightly larger No. 201, from the same catalog, has printed details with no cut-out windows or opening doors, but the second story balcony actually projects from the otherwise flat facade ($700-850). 12.5" high x 9.5" wide x 6" deep. *House and photograph from the collection of Patty Cooper.*

The 1911 Bliss catalog pictures No. 202 with a second story balcony which extends across the front of the house, but this version is the one more commonly found. Unlike the model shown in the catalog, this example has the fishlike "R. Bliss Co" logo, used on earlier houses, printed above the porch entrance. This house was also shown in the enigmatic Butler Brothers advertisement from 1910 reprinted in *Dolls' Houses in America* and like other houses in that photo, the numbers printed on the edge of the papers contradict what might otherwise be a logical progression. Inexplicably, the number 572 can easily be found on at least three different sections of this house. This is the number used in the 1901 catalog for the Keyhole House, although the two houses have no apparent similarities ($1200-1500). 16" high (to top of roof) x 11.5" wide x 7.5" deep. *House and photograph from the collection of Patty Cooper.*

The windows of Bliss No. 203 are printed, but the door is hinged. The 1911 catalog pictures this model with the usual base, which appears to be missing on this example. There are two rooms inside ($1200-1500). 16" high by 13" wide x 7" deep. *House from the collection of Zelma Fink. Photograph by Kathy Garner.*

Although No. 205 is quite a bit larger than the previous houses from the 1911 catalog, it still contains only two long rooms. The windows are printed, but the door is hinged to open. The checked tile that can be seen on the porch floor is the same paper used on the base of the Adirondack shown earlier ($1800-2300). 20" high (to top of roof) x 18" wide x 12" deep. *House and photograph from the collection of Patty Cooper.*

The front of No. 205 opens in one piece. The two long rooms are papered in a small floral print, typical of Bliss houses after 1910, and are large enough to accommodate furniture in 3/4" to the foot scale. *House and photograph from the collection of Patty Cooper.*

American Lithographed Dollhouses and Furniture 17

No. 206 is one of only two houses in the 1911 Bliss catalog with four rooms. The other one, No. 207, is shown on page 11 of *American Dollhouses and Furniture From the 20th Century*. No. 206 is the only 1911 house on which a logo could be found. It is marked with a "B" inside a shield-like shape above the double windows on the second floor. The interior rooms have small patterned wallpaper and do not have connecting doors. The roof has been repainted ($1800-2200). 21.75" high x 20" wide x 11" deep. *House and photograph from the collection of Patty Cooper.*

This house contains a label on the bottom which reads "DIRECTIONS to set up The MOSHER Folding Doll House" followed by detailed instructions and "Patent Applied For." The corners of the house are reinforced with cloth which allows the house to be folded for storage. When set up, it is very sturdy. Although the name of the manufacturer is clearly stated, the house is indistinguishable from the folding model No. 252 shown in the 1911 Bliss catalog. The papers are identical as are the turned columns. Obviously, this is no mere copy, but a later use of the actual Bliss-designed lithographic plates or already printed papers. It is believed to be circa 1920s ($1200-1400). 16" high x 11.5" wide x 9" deep. *House and photograph from the collection of Patty Cooper.*

This unmarked two-story dollhouse has many characteristics of Bliss, although no markings, not even small numbers along the edge, have been found on it. The front pillars are not lithographed as on later Bliss houses nor do they have the same style of turning found on earlier ones, but the hinges are the familiar butterfly shaped ones used by Bliss. Until more catalogs are found, it is impossible to definitely identify the manufacturer of this house, but overall, the full four-color richness of the printing and the whimsical style is reminiscent of Bliss ($1100-1300). 17.5" high x 10.5" wide x 7.5" deep. *House from the collection of Dian Zillner. Photograph by Suzanne Silverthorn.*

Although it is not possible to attribute, with certainty, the later four-room houses to Bliss or Mason & Parker, one other company is known to have used the Bliss papers to manufacture a series of houses. Several dollhouses have been found, clearly marked "MOSHER Folding Doll House/ Patent Applied For," along with set-up instructions. These houses, without a doubt, used the actual lithographic plates or printed papers from earlier Bliss houses. They were not just copies. It is not known exactly when these houses were made, although it has been speculated that they were produced in the 1920s. The most likely explanation is that Mason and Parker, who bought the Bliss company in 1914, later sold either the plates or papers to Mosher for their dollhouses. Because these houses used the actual Bliss designs, they were highly detailed and delightful. They were papered inside with the same small scale prints used by Bliss. Designed to fold for easy storage, they were quite sturdy when set up, with metal rods which both reinforced the house and supported the chimney. Bliss designs known to have been reissued by Mosher include numbers 200 and 202 from the 1911 catalog and the two-room version of the "Adirondack Cabin." Bliss also produced folding dollhouses, including four of the 1911 houses, so the Mosher designs may be indistinguishable except for the labels on the bottom.

Bliss designers used lithographed paper on their dollhouses as an economical way to capture the distilled essence of Victorian architecture in all its shingled, gingerbreaded, finialed glory. By contrast, no attempt was made to use the same medium on the furniture to imitate Victorian styles or details. Instead, the lithographed paper was used only as a surface embellishment in the most whimsical way. Although dollhouse furniture is often seen as a reflection of the design and popular tastes of its time, Bliss furniture is pure fantasy. No Victorian home would have contained chairs adorned with 18" high letters of the alphabet or children wearing pointed hats. The furniture is also chunky and oversized compared to the graceful lines of the houses.

Another house marked "Mosher" is identical to Bliss No. 201, which was shown in the 1911 catalog in both a "regular" version and as folding model No. 251 ($700-850). *House from the collection of Julie Scott. Photograph by Sarah Cox.*

Although unmarked, this folding two-room version of the Adirondack Cabin is believed to have been made by Mosher, circa 1920s. It uses the same lithography, including the deer head gable ornament, as the Bliss cabins. However the Mosher version has the gables on the side instead of the front of the house ($1200-1500). 17" high x 12" wide x 9.5" deep. *House and photograph from the collection of Sharon Unger.*

A side view of the Mosher folding Adirondack Cabin shows the original Bliss lithography complete with deer head. The Mosher version has four sides and is front-opening unlike the Bliss Adirondack which is open-backed. *House and photograph from the collection of Sharon Unger.*

American Lithographed Dollhouses and Furniture 19

The 1896 catalog, reprinted by The Antique Toy Collectors of America, advertised three different sets of "ABC Furniture" as well as three sets of "Parlor Furniture." The drawing on the cover of the ABC furniture set bears no resemblance to any known Bliss house. The 1896 alphabet set was probably the one on which each piece contained letters heavily embellished with a variety of printed curlicues and fruit. It was advertised in three different sizes, No. 264 for 25-cents, No. 265 for 50-cents, and No. 266 for $1.00. It is evident from the furniture which has survived, with corresponding numbers, that these sizes referred to the scale of the furniture rather than the number of pieces in a set. The parlor set was also advertised in three sizes, and various pieces believed to be from each of the three sets are illustrated.

According to Mrs. Jacobs in *Dolls' Houses in America,* the 1901 catalog also advertised furniture in several sizes. From studying boxed sets and marked pieces of furniture, it appears that almost all of the Bliss furniture with illustrations of children was introduced in that year. It is not surprising that the quintessential Bliss furniture would have originated the same year as the most well known Bliss houses. Some of the bedroom pieces which show full figures of children are marked "299" an indication that they are the largest, $1.00 size. Two of the pieces shown in the same photo, the washstand and bureau, are marked "298" which would be the 50-cent size. This mixing of sizes seems to be common among existing sets. There is no way of knowing if this was the original intent or if the existing groupings are the result of packaging leftover pieces from different sets towards the end of the company's life. A different ABC set was advertised in 1901 and is probably the one which includes both pictures of children playing as well as letters of the alphabet. Another series from the same year shows only the heads and shoulders of children, wearing rather exaggerated hats, with a pink and blue background. These are found on a set of parlor furniture as well as a set of "chamber furniture." These pieces are more inexpensively made, with cardboard backs to the chairs and simple turnings on the legs.

The 1895 Bliss catalog shows this bed, washstand, and at least two of the chairs as part of the ten pieces which were included with the purchase of dollhouse No. 240, the "Bow Window" house. The ten piece set probably included a bed, washstand, four chairs, pedestal table, sofa, piano, and bench. The washstand and piano have the number "217" and the other pieces appear to match. This simple piano is the basic Bliss piano, later used with different papers or with the addition of a platform or brackets to make it more compatible with the other furniture designs. The chairs are 3.25" tall and the bed is 6" long (Chairs $95, bed $175, piano $125, washstand $75). *Furniture and photograph from the collection of Patty Cooper.*

Three different size sets of "Parlor Furniture" were listed in the 1896 Bliss catalog. These pieces may have been part of the No. 201 set advertised for 25-cents. The original set would have included a sofa, five chairs, table, and ottoman. The backs of the chairs are made of cardboard and they are 3.25" tall (Chairs $45 each, sofa $85). *Furniture and photograph from the collection of Patty Cooper.*

20 American Lithographed Dollhouses and Furniture

All of these pieces, except for the throne chair in the lower right, are part of set No. 202, advertised as the 50-cent size in the 1896 Bliss catalog. The original set would also have included an ottoman. The chairs are 4" tall (Chairs $85 each, table $95, sofa $150). The throne chair is marked 203 and would have been part of the $1.00 size set, which contained nine pieces. It is 4.25" high ($150). *Furniture and photograph from the collection of Patty Cooper.*

Three different size sets of ABC furniture were advertised in the 1896 Bliss catalog. This set, which does not have pictures of children, was probably the smallest one advertised as the 25-cent size and numbered 264. The sofa and five chairs contain the letters "A" through "S." The piano and bench (not shown) would have been needed to complete the alphabet. The backs of the sofa and chairs are cardboard. The chairs are 3.5" tall (Set of sofa, five chairs, and table $600). *Furniture and photograph from the collection of Patty Cooper.*

The cover of a boxed set of Bliss No. 265 furniture, advertised in the 1896 catalog for 50 cents, shows some of the same designs used on the lithography of the chairs. The furniture depicted bears little resemblance to the actual pieces nor does the dollhouse look like anything ever produced by Bliss. The full set contains five chairs, a sofa, checkerboard table, floor lamp, jardiniere, and piano with bench (Boxed set $1600-2000). *Boxed set and photograph from the collection of Carl and Linda Thomas.*

American Lithographed Dollhouses and Furniture 21

Some pieces from set No. 265 are shown. Three chairs are missing from this group, making the alphabet incomplete. The chairs are 4.75" tall and the backs are made of wood. The piano is the basic one shown as part of set No. 217, but it has been embellished by the use of alphabet paper and enlarged by the addition of wood pieces to form a platform, piano lid, and brackets. The small piece on the lower left was probably intended to be a jardiniere. Resembling a chess piece with two lithographed cherries on the top, it is difficult to identify, out of context, as dollhouse furniture, much less as Bliss (Chairs $85 each, sofa $175, table $110, piano $200-250, jardiniere $45). *Furniture and photograph from the collection of Patty Cooper.*

One piece from set No. 266, the $1.00 size in the 1896 catalog, is shown on the right to provide a comparison with chairs from the smaller sets. It is 5.75" tall. The original set would have included five chairs, a table with checkerboard top, a sofa, a lamp, an armchair, and a piano with bench (Individual chair $95). *Furniture and photograph from the collection of Patty Cooper.*

The same series of numbers were used on another, probably later, ABC parlor set. These 25-cent size pieces are the same size and shape as those shown previously as No. 264, but the lithography has been redesigned to include drawings of children along with the alphabet. The chairs have cardboard backs and are 3.5" tall. This set contains the complete alphabet ($750-800). *Furniture and photograph from the collection of Ray and Gail Carey.*

The medium-size set with both the alphabet and pictures of children is, like its 1896 predecessor, marked "265". The chairs have wood backs and are 5" tall. It is not known if the "alphabet with children" motif was ever produced in the large $1.00 size (Chairs $85, sofa $175, table $110). *Furniture and photograph from the collection of Patty Cooper.*

Also advertised in the 1901 catalog were several sets of furniture depicting children with hats. The 25-cent size was No. 493, a parlor set which was said to have contained nine pieces. The three-legged chairs have cardboard backs and are 4" tall (Sofa $150, chairs $85, table $85, lamp $75). *Furniture and photograph from the collection of Patty Cooper.*

This Bliss bedroom set has lithographed pictures of children which are very similar to those found on the later alphabet sets, suggesting that both may have been made around the same time. The washstand and bureau of this group are marked "298", the number used in the 1901 catalog for the 50-cent size bedroom set. It is possible that the bed is actually from set No. 299, the $1.00 size. The chairs are 5" tall and the bed is 8.5" long (Set $800-1000). *Furniture and photograph from the collection of Patty Cooper.*

The cover of this boxed set helpfully displays the number 493 in the lower right hand corner and the Bliss logo, the same one used on the 1901 houses, in the lower left. The set contains five chairs, a sofa, table, and lamp (Boxed set $1200-1500). *Boxed set and photo from the collection of Carl and Linda Thomas.*

The matching bedroom, No. 494, was advertised in the 1901 catalog as containing seven pieces. Most likely, the set included a bed, table, lamp, bureau, rocking chair, and two side chairs. The bed is 6.5" long (Bed $175, bureau $150, table $85, lamp $75, rocking chair $100). *Furniture and photograph from the collection of Patty Cooper.*

A SERIES OF LITHOGRAPHED HOUSES BY AN UNKNOWN MAKER

There are a total of six models known to be part of this series of lithographed dollhouses produced by an as yet unidentified company. Lithographed toy collector, Bill Dohm, coined the unflattering, but descriptive term "Gutter Houses" when he noted that most of the houses were adorned with a piece of over scale molding where the eaves troughs, or gutters, would be on a full size house. It is obvious from their architectural style and interior papers that the houses were produced some time around the beginning of the twentieth century and are contemporaries of Bliss. It also seems clear that they were manufactured in America because that is where they have been found. The houses range in size from a tiny two-room house, only 10.25" tall, to a six-room apartment building almost 27" tall.

Often misidentified as Bliss or Whitney S. Reed, the houses bear characteristics which distinguish them from dollhouses manufactured by either of these two companies. One of the most obvious is the prominent molding from which their nickname is derived. In addition, the hinges which allow the fronts to open are long and horizontal, unlike the butterfly shaped hinges used by Bliss and Reed. The range of colors used in the lithography is much narrower than on Bliss houses. On the Gutter houses, the colors are mostly limited to golden brown, forest green, and red. All six houses have the same design for the door which features a lace-curtained window, below which are three horizontal panels and three vertical panels. The letters or numbers often found on the edges of Bliss or Reed houses are absent on these unidentified houses. On five of the Gutter houses, the papers which wrap around the edges are treated as contrasting corner boards printed in green. This is quite different from Bliss and Reed houses on which the pattern of brick, stone, or clapboard continues around the edge. The same five Gutter houses have identical bands of horizontal fish-scale shingles decorating the second story exteriors. The sixth house has a lithographed stone pattern on the first floor with a simple brick pattern on the second floor, quite similar to the early Reed houses. The "stone" and "brick" on this house defy the rule and wrap around the edge as on a Bliss, but it is clear that this

Although the name of the company which made this series of houses is unknown, they may be grouped under the nickname Gutter Houses because of the over scale molding around the roof edges. The smallest house contains two rooms and is 10.25" tall x 8" wide x 4" deep ($500-600). *House and photograph from the collection of Patty Cooper.*

house, with its round turret, is still part of the Gutter family. It has the same turned columns, same door, same hipped roof, and same paper around the base as is found on other houses in the series.

Inside, the Gutter houses are papered in a very Victorian, over scale floral pattern, which appears to be full-size wallpaper. Similar papers were used in older Bliss houses, but early in the twentieth century, Bliss began using rather simple, patterned wallpapers in a more appropriate dollhouse scale.

It is the wallpaper which provides a fragile link between the known Gutter houses and a seventh house which, although it contains almost none of the other identifying features, may also have been manufactured by the same unknown company. Collectors have long noted the variations among different examples of the lithographed log cabins attributed to Bliss. The most recognizable difference is that some have a deer head printed in the front gable while others have only a log pattern. Closer examination reveals that, in fact, the lithography on the two types is completely different, log by log, line by line. There is no logical reason why Bliss would have gone to the expense of making two different sets of plates for such similar houses. It is much more likely that one company copied the other. There are several clues which indicate that the cabin with deer head is the Bliss version. Therefore, the one without the deer head must have been made by another company, perhaps the one which made the Gutter houses. This theory is reinforced by the fact that the wallpaper usually found in the "deer-less" versions is an over scale floral—not very cabin-like, but quite typical of that used in the Gutter houses. Several other Gutter houses bear marked resemblances to known Bliss houses, so it seems likely that at least one of the companies was keenly aware of the other and not above taking their inspiration from the competition. None of the known Gutter houses is post-Victorian in style, so either their dollhouses changed beyond recognition or the company ceased production by the first World War.

This house bears a marked similarity to Bliss house No. 273, the Bay Window House. The two houses are nearly identical in size ($1200-1400). 18" tall x 10.25" wide x 9" deep. *House and photograph from the collection of Patty Cooper.*

This slightly larger, two-room house displays the three colors most often used on Gutter houses ($700-800). 13" high x 9.75" wide x 4.75" deep. *House and photograph from the collection of Patty Cooper.*

American Lithographed Dollhouses and Furniture 25

The brick and stone pattern on the exterior papers are atypical of Gutter houses, but other features such as the door, base papers, columns, and hinges indicate that it is part of this series. The beading pattern printed around the door and windows is also identical to that found on other Gutter houses. The one story turret is an unusual feature which increases the value of this house ($1800-2000). 18.5" tall x 12.5" wide x 10" deep. *House and photograph from the collection of Patty Cooper.*

The two-room interior shows the over scale floral wallpaper usually found in Gutter houses. The temporary attic floor has been added to provide a bathroom. As designed, the upstairs room had a "cathedral" ceiling. *House and photograph from the collection of Patty Cooper.*

The simple, two-room interior is rather disappointing compared to the architectural exuberance of the exterior. The over scale wallpaper is typical of this series. *House and photograph from the collection of Patty Cooper.*

Rather thin evidence suggests this log cabin might actually be part of the Gutter house family. Although it is difficult to see in a photograph, every detail of the lithography is different from the papers on the Bliss Adirondack cabin. Differences which are more easily observed are the opposite slant of the roof, the absence of the deer head in the gable, and a distinctly different log pattern on the balcony railing. This house has symbols penciled on the bottom which may someday provide a clue to its origin: B+=Gq above a line with 1.89 below ($1200-1400). 17.25" tall x 18" wide x 9.5" deep. *House from the collection of Madeline Large. Photograph by Patty Cooper.*

Another two room house features a walled garden, two-story bay window, and a hipped roof. Almost identical paper was used on the largest version, a three-story apartment building shown in *Dolls' Houses in America* by Flora Gill Jacobs. The paper on the base of this house has been painted over, but remnants show that originally it was the same as the base paper on the house with the turret. Two finials are also missing from the garden wall ($1200-1400 in condition shown). 22" tall x 14.5" wide x 12" deep. *House and photograph from the collection of Patty Cooper.*

The interior of the cabin is papered with an over scale floral pattern typical of Gutter houses. The four rooms are all accessible from the back, unlike the Bliss version in which the attic room is accessible by lifting a printed "Indian head" on the roof of the dormer. *House from the collection of Madeline Large. Photograph by Patty Cooper.*

Like their Bliss contemporaries, the log cabins in this series came in both two and four-room versions ($800-1000). The rather unlikely colored red and blue isinglass windows have also been found on Bliss Adirondack cabins. 17.25" tall x 12" wide x 8.25" deep. *House and Photograph from the collection of Patty Cooper.*

WHITNEY S. REED COMPANY

The Whitney S. Reed Toy Company was founded in Leominster, Massachusetts, in 1875. Primarily a manufacturer of other types of wooden toys, the company also made dollhouses and some dollhouse furniture. The company marked many of their toys "W.S. Reed Toy Company" with a patent date, thus guaranteeing a certain fame among today's toy collectors. Unfortunately, this was not the case with their dollhouses and this lack of marking has contributed to much confusion in identifying Reed dollhouses. In 1897, the company was sold and eventually renamed the Whitney Reed Chair Company, although they continued to manufacture toys.

W.S. Reed made dollhouses in a variety of sizes and styles. During the 1890s, the houses were small in scale with detailed lithographed exteriors. These houses were typical of the Italianate style popular in America in the mid-1800s. They were mostly symmetrical and sometimes boasted square towers. Many of the houses had windows with rounded tops and doors with elaborate, often pedimented crowns, consistent with the Italianate style. This design helps to distinguish them from Bliss houses which were mostly Queen Anne in style. However, the best clues to identifying a Reed house are a series of small printed "x"s used to facilitate the placement of the appropriate paper on each house. On the smallest example, "x2" is printed on the side and "x3" on the gable. On the next larger size, "xx2" is printed on the side, "xx3" on the gable, and "xxxx1" on the door.

Perhaps because Flora Gill Jacobs described them in her book *Dollhouses in America* as the "Vanished houses of Whitney Reed," these houses have acquired a certain mystique among dealers and collectors. Although several Reed houses have been added to the Washington Dolls' House & Toy Museum collection since the publication of her book in 1974, they are still perceived to be rarer than other lithographed paper over wood dollhouses. This perception, combined with the fact that collectors of other toys by Reed also seek the dollhouses, usually allows them to command a slightly higher price than similar dollhouses from the same period. However, collectors should exercise caution when searching for a Reed houses. Because little has been written about them, any house that cannot otherwise be identified is often attributed to Whitney Reed.

Conversely, houses that actually are Whitney Reed are frequently misidentified. The early Reed houses, with their lithographed exteriors and blue roofs, are easily mistaken for Gottschalk Blue Roofs. To further complicate matters, the Reed houses are very similar to the early paper over wood dollhouses manufactured by Converse before that company began printing their designs directly on wood. Converse was located in Winchendon, Massachusetts, a town less than thirty miles from Leominster. The proximity of the two companies and the similarity of their early designs suggests that perhaps their papers were obtained from the same source. Based on examination of a limited number of examples, it appears that the houses of the two companies can be differentiated by the shape of the "stonework" on the first floor. The Converse houses have smoothly rounded "stones" unlike the ones on the Reed houses which are either squared-off or very irregular in shape. On at least one example, the squared tower of a Converse house is on the far right while the known Reed houses are symmetrical. On all of the Reed houses, except the smallest, at least some of the windows are cut out whereas none of the windows are cut out on the lithographed paper over wood houses known to be Converse.

This two-room house by Whitney S. Reed, circa 1897, was no longer shown in their catalog after the turn of the century ($800-1000). 8.5" high x 5" wide x 4" deep. *House and photograph from the collection of Patty Cooper.*

28 American Lithographed Dollhouses and Furniture

After the turn of the century, the Reed lithographed houses began to be superseded by one-story houses on which the exterior papers showed a brick pattern but no other architectural details. Instead, actual wood pieces were used for the frames of the doors and windows. These "New Practical Doll Houses" were open-backed and often featured a "bull's eye" decoration centered in the gables. They ranged in size from one to three rooms. The lithographed doors were usually green with yellow shades and lace-edged curtains in the printed windows. On the known examples, the tell-tale Reed "x"s are printed on the doors.

The front of the house opens to reveal two simple rooms with small-patterned wallpapers. The floor paper on the first floor has been replaced. *House and photograph from the collection of Patty Cooper.*

Only slightly larger, this circa 1898 house features cut out windows and an opening door. Curtains are printed on the first floor windows but made of real lace on the second floor. It has an unusual sleigh type base ($1000-1200). 11.5" high x 7.5" wide x 7" deep (at base). *House and photograph from the collection of Patty Cooper.*

Another lithographed Reed house from the same period features a central tower. 14" high x 8.25" wide x 6.5" deep ($1200-1600). *House from the Washington Dolls House & Toy Museum. Photograph by Suzanne Silverthorn.* $1200-1600.

An example of the dollhouses introduced by Whitney S. Reed circa 1902, this one room cottage is open-backed. The lithography is a simple brick pattern and architectural details are applied with actual wood pieces. Note that the door is the same as on the preceding house ($300-500). 8.25" high x 8.5" wide x 5" deep. *House and photograph from the collection of Rita Goranson.*

DUNHAM'S COCOANUT

The Dunham's Cocoanut dollhouse dates from the 1890s when it was used to ship shredded coconut to grocery stores. When emptied, the wood crate could be set in a vertical position and converted to a four-room, open fronted dollhouse. Lithographed paper was used to line the inside of the house, creating four brightly colored and highly detailed rooms which exemplify Victorian decor. No one knows for sure if the crates arrived with or without the paper glued in. However, all known examples show the same configuration of rooms, with the bedroom at the top, the parlor below, dining room on the second story, and a kitchen at the bottom. This suggests that the paper must already have been in place or else detailed instructions were provided and everyone followed them. All of the rooms had 7" high ceilings.

The exterior of the houses had minimal architectural interest. Each side was usually covered with paper printed in black with a rather large brick pattern, four windows with slightly rounded pediments, and the words "trade mark registered." The ends of the crate, which became the bottom and top of the dollhouse, were printed with the name "Dunham's Cocoanut Doll House." The backs of the houses were plain and most of the ones found today usually contain a crack, not surprising given the inexpensive lumber typically used for crates. The dollhouse was 29" high, 11.75" wide, and 7.25" deep.

The bedroom of the Dunham's Cocoanut dollhouse was on the top floor. It featured blue-striped wallpaper with a border near the ceiling. There is an oriental rug on the floor. A lace-curtained bay window on the back wall has stained-glass panels. The ceilings are 7" high. *House and photograph from the collection of Patty Cooper.*

Below the bedroom, the parlor shows typical Victorian interior decoration. The upright piano holds sheet music and photographs. There are paintings on the walls and the requisite ferns, one on a tripod table. *House and photograph from the collection of Patty Cooper.*

30 American Lithographed Dollhouses and Furniture

Cardboard furniture to furnish all four rooms could be ordered from the Dunham company, which was originally located in St. Louis but later moved to New York. The original furniture for the bedroom included a bed, chair, mirrored bureau, and commode. The parlor pieces included a sofa and two arm chairs with matching ottomans. Two arm chairs, a table, and a mirrored buffet with a chafing dish would have furnished the dining room. The kitchen pieces consisted of a stove, table with cooking utensils, and two wooden chairs. This furniture was very fragile and is nearly impossible for today's collector to find. Most of the pieces were printed with the trademark cake picture and the name "Dunham's Cocoanut," making it easy to identify in the unlikely event that a piece is encountered.

The second-story dining room has a chair rail with red wallpaper below and blue above. An aquarium is in front of the bay window whose sill holds several house plants. There is a moose head on the wall and an oriental rug on the floor. *House and photograph from the collection of Patty Cooper.*

The kitchen, at the bottom of the house, has a red and white checked floor and tongue-in-groove wainscoting. A built-in cabinet, on the left wall, holds dishes and several boxes of Dunham's Cocoanut. The back wall contains an hourglass, a clock, and a folding drying rack. Since many dollhouse decorators choose to locate a stove on this wall, the drying rack is often mistaken for a stovepipe in photographs. *House and photograph from the collection of Patty Cooper.*

This Dunham's Cocoanut house shows the original cardboard furniture which could be ordered from the company by sending in trademarks from the coconut packages. The furniture is easily identified by the trademark coconut cake with the Dunham's Cocoanut name printed on most of the pieces. 29" high x 11.75" wide x 7.25" deep ($800-1200 house, furniture is too rare to establish price). *Photograph by D. Dow.*

MCLOUGHLIN BROTHERS

The McLoughlin Brothers company of New York was famous for producing brightly lithographed books, games, and toys beginning in the late 1850s. The company started making dollhouses as early as 1875 (see *American Dollhouses and Furniture from the Twentieth Century.*) McLoughlin is generally thought of as a manufacturer of folding cardboard dollhouses, but at least one of their early dollhouses was made of wood. "Dolly's Play House," shown in their 1875-76 catalog, was open-fronted and contained two rooms with elaborate interior lithography. The sides of the house were covered in a plain brick paper with no architectural features. However, a lavishly decorated cornice frames the front of the house, giving it the appearance of a stage set. This same house was available later as a folding cardboard model.

One of the best known McLoughlin dollhouses is actually a set of folding rooms patented in 1894. It was made of heavy cardboard and folded into a flat package, measuring 12" x 12" and approximately 1" thick. The cover contained directions for unfolding the house to reveal a kitchen, dining room, bedroom, and parlor, each with two walls and a floor elaborately printed to show typical Victorian decor. A cut-out, draped archway provides access from the parlor to the dining room. The opened house was two feet wide and two feet deep, easily accommodating furniture of 1" to the foot in scale.

McLoughlin also produced at least two sizes of paper furniture for their houses. The set shown here is 1/2" to the foot in scale, too small for the folding rooms, but fitting quite nicely in "Dolly's Play House" or the "Garden House" shown in the previous volume. Although the uncut sheets of furniture carry a much greater monetary value, it is still a joy for collectors to find assembled pieces with which to furnish a McLoughlin house. Carefully cut out by young hands, the fact that any of the fragile pieces have survived is evidence that a large quantity must have been printed. Known sets included a kitchen, parlor, Mission-style library, dining room, and several bedrooms. Each contained many accessories such as pictures, lamps, and vases. The McLoughlin furniture is easy to identify because many of the pieces contained printed mottoes. The top of the china cabinet bears the inscription "Haste Makes Waste" and the front of the icebox proclaims "No Sauce Like Appetite." A cat and a strange vegetable creature labeled "The Chief Cook" are part of the whimsical kitchen set which also has a lobster adorning the table. At least two of the pieces, the piano and a vegetable slicer, are printed with the McLoughlin name.

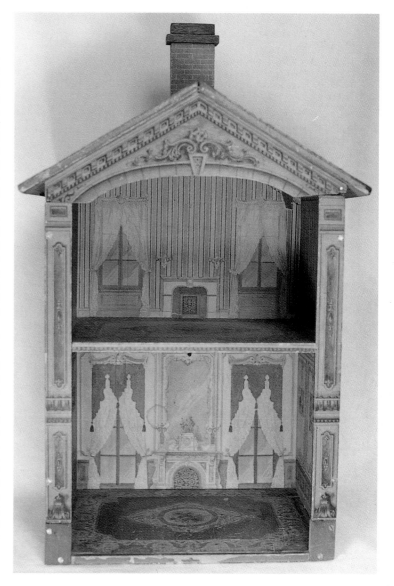

An early "Dolly's Play House," circa 1875-85, was constructed of wood. Later folding versions were made of cardboard ($1500-1800). 18" high (to peak of roof) x 13.5" wide x 10" deep. *House and photograph from the collection of Patty Cooper.*

The parlor of the McLoughlin folding table-top dollhouse features an elaborate fireplace and draped archway leading to the dining room ($700-850). Room is 12" square. Unfolded house measures 12" high x 24" wide x 24" deep. *House and photograph from the collection of Patty Cooper.*

The McLoughlin dining room features an oriental rug on the floor and a deer head over the fireplace. *House and photograph from the collection of Patty Cooper.*

The well-equipped McLoughlin kitchen includes a coal or wood burning stove with water heater, a drop-front cabinet, and a plumbed sink. *House and photograph from the collection of Patty Cooper.*

The cozy McLoughlin bedroom displays the motto "Home Sweet Home." *House and photograph from the collection of Patty Cooper.*

American Lithographed Dollhouses and Furniture 33

This McLoughlin furniture is in approximately 1/2" to the foot scale but it is likely that the same designs were used for larger scaled furniture. The parlor pieces originally included two tables, a floor lamp, pictures, and vases in addition to the pieces shown here (Cut pieces $4-6 each). *Furniture and photograph from the collection of Patty Cooper.*

McLoughlin produced at least two different bedroom designs. This photograph probably contains some pieces from each as well as the chaise lounge from the dining room set. The bedroom sets would have also included lamps, rugs, and items to hang on the wall (Cut pieces $4-6 each). *Furniture and photograph from the collection of Patty Cooper.*

The McLoughlin furniture can be identified by the printed mottoes such as "Haste Makes Waste" on this china cabinet. The original dining room set also included lamps, vases, and the chaise lounge mistakenly shown in the bedroom photo (Cut pieces $4-6 each). *Furniture and photograph from the collection of Patty Cooper.*

The detailed McLoughlin kitchen originally included several kitchen utensils, a cat, and a vegetable creature. The lid of the dry sink proclaims "Water Water Everywhere" and the icebox exhorts "No Sauce Like Appetite." (Cut pieces $4-6 each). *Furniture and photograph from the collection of Patty Cooper.*

Each sheet of McLoughlin furniture originally contained a variety of accessories. The kitchen slicer, to the left of the "Letters" holder, contains the McLoughlin name. The vegetable creature in the middle of the photo is "The Chief Cook" and belongs in the kitchen (Cut pieces $4-6 each). *Furniture and photograph from the collection of Patty Cooper.*

F. CAIRO

In her book *A History of Dolls' Houses*, Flora Gill Jacobs shows an uncut sheet of furniture made by "F. Cairo of Brooklyn in 1892." Her sheet is labeled "Our Dining-room" and the text states that Mrs. Jacobs also has the matching "Our Parlor" and "Our Bedroom" sets. (It is not known if "Our Kitchen" was also made.) Were it not for this helpful information, it would be easy to assume that the unmarked, heavy paper furniture was made by McLoughlin. Both sets are from the same time period, in the same style, and designed to be assembled in the same way. The main difference between the two sets of furniture is the size. The F. Cairo furniture is in approximately 3/4" to the foot scale while the McLoughlin pieces shown are 1/2" to the foot. However, since McLoughlin is said to have made larger furniture, this distinction may be of little help to the collector. The similarity suggests that both sets, although manufactured by different companies, may have been designed by the same hand or, at the very least, that one company was strongly influenced by (i.e. copied) the other. Aside from minor decorative variations which would be difficult to note without comparing piece by piece, the most significant difference between the two sets appears to be the absence of mottoes on the F. Cairo furniture. As it is difficult to find paper furniture in good condition, the F. Cairo furniture would be a good substitute in a Dunham's Cocoanut house or in the largest McLoughlin. Paper furniture is usually so fragile and so rare that the collector who wishes to furnish a house should consider using reproductions, made by color copying, and storing the originals flat, between sheets of acid-free paper.

"Our Bedroom" pieces by F. Cairo circa 1892 are made of heavy paper and very fragile. The original sheets would have also included accessories. The pieces shown are color copies as the owner felt the originals were too fragile to use or photograph (Original cut pieces $10-12 each). *Furniture and photograph from the collection of Patty Cooper.*

The F. Cairo company of Brooklyn produced heavy paper furniture in approximately 3/4" scale circa 1892. The sideboard and chaise lounge were originally from the "Our Dining Room" sheet. The other pieces were probably part of "Our Parlor" (Original cut pieces $10-12 each). *Furniture and photograph from the collection of Patty Cooper.*

GRIMM & LEEDS

The ability to fold was apparently a desirable quality in a turn-of-the-century dollhouse, as evidenced by the number of companies who made them, e.g. Bliss, McLoughlin, and Stirn & Lyon. It is possible that there were other designs, but the only dollhouses known to have been made by Grimm & Leeds were collapsible. Their cardboard houses were cleverly designed to be assembled from pieces which packed flat except for the porch roof and the box which formed the foundation of the house. When assembled, the houses were surprisingly sturdy. Grimm & Leeds was one of those manufacturers which delight collectors by helpfully printing their name on their dollhouses. Inside the foundation box, a label was pasted which contained the information: "Leeds Toy House, manufactured by Grimm & Leeds Co., Camden, N.J. Patented Sept. 23d, 1903. Four varieties." From the two examples shown here and one pictured in Flora Gill Jacobs' *Dolls' Houses in America*, it is apparent that the four varieties consisted of the red brick "Colonial" in two and four-room versions and the "Dandy Toy House" with a combination fish-scale shingles and stone, also available in both sizes. All of the houses were in the same scale, but the four-room houses were twice as wide as the two-room models. The insides of the houses were decorated in a small leaf pattern, with the same design on both the upstairs and downstairs walls. The cardboard floors were printed with a wood-grain pattern as were the operable, windowed doors. Access to the houses was through the open backs. The cut-out windows contained isinglass panes and paper shades with lace trim. In the two examples shown, the windows are, amazingly, still intact.

A similar two-room Grimm & Leeds "Dandy Toy House" features fish scales "shingles" and "stone" printed on the exterior ($500-600). 20.5" high x 9.5" wide x 14" deep. *House and photograph from the collection of Patty Cooper.*

The Grimm & Leeds dollhouses were available in both two- and four-room versions. This two-room "Colonial" model has a printed "brick" exterior ($500-600). 20.5" high x 9.5" wide x 14" deep. *House and photograph from the collection of Patty Cooper.*

The open back provides access to the two-room interior of the Grimm & Leeds "Dandy Toy House." The walls are printed in a small leaf pattern and the windows still have their original lace-edged green shades. *House and photograph from the collection of Patty Cooper.*

ENGLISH DOLLHOUSES AND FURNITURE

For many people the ideal of an antique dollhouse is best exemplified by those produced in England. We picture nurseries with a nanny, a window seat, and a lovely old dollhouse in the corner. To this day, the British seem to have a love of miniatures surpassed by none. This is evidenced by the number of high quality magazines devoted to the hobby which are being published in England. Although, as is true of such magazines in the United States, most of the pages are devoted to new miniatures, the British magazines usually contain at least one article of interest to the collector. (See list of resources at the end of the book for magazines titles and addresses.)

G. & J. LINES AND LINES BROTHERS

The largest and best known manufacturers of dollhouses in England were the related firms of G. & J. Lines and Lines Brothers. British researcher, Marion Osborne, has published a book which contains reproductions from the catalogs of both companies. Her book, *Lines and Tri-ang Dollshouses and Furniture 1900-1971*, provides the most comprehensive source of information available on these companies. The original brothers, George and Joseph Lines, are believed to have begun making dollhouses in the 1890s. Their dollhouses ranged in size from a simple, flat-fronted, box-like structure less than 15" high to an elaborate three-story townhouse which was 43" tall. Many of their dollhouses had lithographed "brick" paper decorating the exteriors. One series, known to collectors as "Kit's Coty" houses, featured bay windows and elaborate roof lines with a widow's walk and multiple chimneys. G. & J. Lines made few changes in their designs, continuing to produce houses which were late Victorian in style well into the 1920s. This makes it very difficult to establish an accurate date for a Lines dollhouse. Often they are thought to be much older than they really are.

The G. & J. Lines Company did not manufacture furniture for their dollhouses. Apparently they found it unnecessary or uneconomical to compete with the high quality furniture which could be inexpensively imported from Germany at the time.

Three of Joseph's sons established their own company soon after World War I. The new Lines Brothers firm coexisted with the earlier company for over a decade until the two merged in 1931. The three Lines brothers used the trade name Triangtois, later changed to Tri-ang. Reflecting styles popular in England at the time, many of the Tri-ang designs were variations on the Tudor theme. Sizes ranged from the tiny "DH/A" which stood 19.5" tall to the "Perfect Doll's Countryhouse" which was an impressive 43" tall and 70" long. It contained eight rooms and came with its own folding stand. All of the houses were completely finished with floor papers and wallpapers. Fireplaces were provided in all the houses and many of the early houses contained cooking ranges. The larger houses had staircases. The windows were made of metal and had curtains, with opening windows first appearing around 1930. Many of the houses came with electric lighting and spaces provided for batteries. The lighting feature was eliminated during World War II and not provided again until the late 1950s. The doors were functional and ornamented by a trademark brass knocker and mail slot embossed with the word "letters." The larger houses also had fitted bathrooms. In the 1920s, the largest Tri-ang dollhouse included a garage and by 1939, several of their more modest dollhouses contained this feature.

In 1932, Princess Elizabeth was given a child-sized Welsh cottage for her 6th birthday. The house was displayed in London and inspired Tri-ang to manufacture dollhouse versions in several sizes. Both front and back

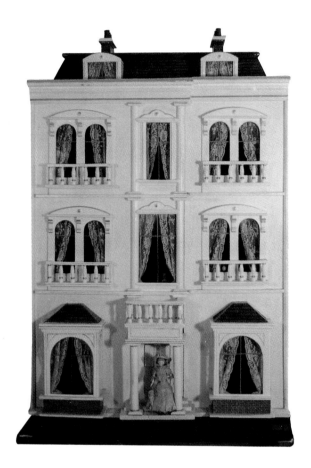

This large house by the firm of G. & J. Lines contains six rooms and three hallways with a central staircase. It is shown in their 1909/10 catalog as No. 16, but was probably available many years earlier. The house could be purchased with a fitted bath and the catalog suggested that a tank could be hidden under the lift-up gable roof (House is too rare to determine price). 52.5" high x 33" wide. *House and photograph from the collection of Anne B. Timpson.*

opening models were available, some with garages. The houses had textured "stucco" exteriors and faux thatched roofs. At least one model of the Tri-ang Princess dollhouse was available as late as 1957.

In 1939, Tri-ang introduced their Modern Flat Roof style dollhouses in five different sizes. The two largest were scaled for 1" furniture and the others accommodated furniture that was 3/4" to the foot. The houses had a textured "stucco" finish and all but the smallest included a movable suntrap on the roof. Unfortunately, this suntrap is often missing when the houses are found today. The houses featured covered porches, garages, and electric lighting.

The popular Tudor styles were included in the Tri-ang line through the late 1950s, making it rather difficult to date these houses. Perhaps most

English Dollhouses and Furniture 37

Advertised as house No. 20 in the circa 1910 catalog of G. & J. Lines, this house opens in two sections to reveal four rooms. It contains a stairway and four fireplaces (House to rare to determine price). 30" high x 24" wide. *Private Collection.*

some Tri-ang models had metal fronts. During the 1960s, several styles were produced in fiberboard with plastic roofs, windows, and door frames. Most of the houses were very contemporary in style.

The Lines Brothers Company was one of those manufacturers, dear to the hearts of purist dollhouse decorators, which also produced furniture. Unfortunately, most of the early furniture was in a small 1" scale which only fit into their larger dollhouses and is extremely hard to find in the United States. In the early 1920s, much of the Lines furniture was manufactured by the Eric Elgin firm of Enfield. This sturdy furniture was made of wood, with a dark stain, and had opening doors and drawers with metal pulls. The style was described as "Jacobean" and many of the pieces were decorated with impressed designs. Even after the Elgin factory closed in 1926, Lines Brothers continued to make furniture using Eric Elgin's original designs. By the 1930s, Tri-ang introduced a new line of dollhouse furniture in Queen Anne and Jacobean styles. This furniture also included pieces upholstered in small scale chintz patterns. An unusual nursery set was decorated with Walt Disney characters. Tri-ang furniture was sold under the trade name "Period Doll Furniture."

commonly found of the Tudor style houses are those in the series numbered 60 through 64. These houses were all approximately 16.5" tall, but varied in width from No. 60 which was 13.75" wide, with only one gable and side porch, to No. 64 which was 45" wide and contained three gables and a garage. All of these houses were appropriate for small scale, 1/2" to 3/4" furniture, such as that produced by Dol-Toi. They were available from the early 1930s through the late 1950s, with only minor changes. At least two versions of this house were also produced by Lines Brothers in Canada under the trade name Thistle. A Thistle catalog from 1950 shows the two smaller sizes, very similar to the Tri-ang models, but opening from the back.

Another popular Tudor dollhouse was the larger scaled model known to collectors as the Stockbroker Tudor, but listed in the Tri-ang catalogs as No. 93. The house, which easily accommodated the 1" scale furniture produced by Tri-ang, opened in four sections. It had a garage with double doors and a built-in bath which included a bathtub, lavatory, toilet, medicine cabinet, and towel rack. The kitchen contained a Welsh dresser, gas cooker, and sink. There was a central stairway with a landing. The living room and each of the two bedrooms contained a fireplace. On the right side of the house was a small service porch and an imposing chimney which featured a sundial.

Tri-ang dollhouses from the 1950s and 1960s are encountered much more rarely in the United States than one might expect. By the late 1950s,

The kitchen of Lines No. 20 is located on the lower left. It contains the original wallpaper and floorpaper, as well as a Lines fireplace and kitchen dresser. *Private Collection.*

38 English Dollhouses and Furniture

When World War II began, Tri-ang stopped making dollhouse furniture. The cost of producing small wood pieces was said to be prohibitive. However, in 1960 the company set up a factory in Belfast to manufacture die cast plastic pieces. This furniture was in 3/4" to the foot scale, much smaller than the early furniture, and very contemporary in design. It was marketed under the trade name "Spot-On." A wide range of accessories were included in the line. Apparently, the Spot-On line did not do very well. In 1965, Tri-ang redesigned the furniture and changed the name to Jennys Home. In addition to furniture and accessories, the Jennys Home line offered a family of dolls and individual rooms which could be combined to create a dollhouse. Tri-ang went out of business in 1971. Some of the Jennys Home designs were used for the Blue Box furniture, manufactured in Hong Kong and sold through mail order catalogs in the United States.

The inside of the "Kit's Coty" type house contains four rooms with fireplaces. The stairway has been replaced. *House from the collection of Dian Zillner. Photograph by Suzanne Silverthorn.*

One of series known to collectors as "Kit's Coty", this large house was advertised by the G. & J. Lines company circa 1910. It features two chimneys with chimney pots, a widow's walk, and two story bay windows ($2000-2500). 32" high x 24.5" wide x 17" deep. *House from the collection of Dian Zillner. Photograph by Suzanne Silverthorn.*

The 1915 G. & J. Lines catalog showed this early version of No. 35, with an intricate decoration on the central section. The glass windows featured printed blinds (House too rare to determine price). 22" high x 19" wide x 11" deep. *Private Collection.*

Inside the early version of No. 35 are two long rooms with the original fireplaces. *Private Collection.*

G. & J. Lines advertised model No. 35 in their 1925 catalog as "improved this season" and the changes from the earlier model are evident. The opening front is a combination textured finish (described as "roughcast" in the catalog) and lithographed brick paper. The windows had glass panes with blinds and curtains. The two long interior rooms have ceiling heights of 5.75" ($900-1000). 20.5" high x 18.75" wide x 10.75" deep. *House and photograph from the collection of Ray and Gail Carey.*

One of the smallest of the early Triangtois dollhouses, Cottage No. DH/B was advertised in their 1921 catalog. It is very similar to No. 22 sold over a decade earlier by the parent firm of G. & J. Lines ($700-800). 21.5" high (not including chimney) x 14" wide x 7.5" deep. *House and photograph from the collection of Ray and Gail Carey.*

The front of DH/8 opens in one section to reveal four rooms and a stairway. The house has its original wallpapers and fireplaces. The kitchen contains a Tri-ang Welsh dresser and kitchen range. *House from the collection of Dian Zillner. Photograph by Suzanne Silverthorn.*

This medium-sized dollhouse was shown in the 1921 Triangtois catalog as Cottage No. DH/8. Its exterior was "rough cast" on the second story and papered in "brick" on the first. In the 1924 catalog, this number was used to designate another dollhouse, described as a Devonshire Cottage, which was approximately the same size and shape, but with brick paper on both the first and second stories. The fence and gate were not shown in the catalog ($900-1000). 27" high x 24.5" wide x 14.5" deep. *House from the collection of Dian Zillner. Photograph by Suzanne Silverthorn.*

A slightly larger house, described as Surrey Cottage No. DH/10 in the 1921 Triangtois catalog, features an unusual entrance ($1200-1400). 32" high x 31" wide x 16.5" deep. *House and photograph from the collection of Ray and Gail Carey.*

The front of Surrey Cottage opens in two sections and contains four rooms with a central hall with stairway. The house has its original wallpapers, fireplaces, and kitchen dresser. Note the built-in kitchen range which differs from the one shown earlier. The house is furnished with German red-stained furniture. *House and photograph from the collection of Ray and Gail Carey.*

English Dollhouses and Furniture 41

In the 1924 Triangtois catalog, the number DH/10 was used for a slightly different house which featured half-timbering on the second story and a redesigned front entrance. This house was approximately the same size as the earlier DH/10 and had the same configuration of rooms inside ($1200-1400). *House and photograph from the collection of Leslie and Joanne Payne.*

The second largest dollhouse in the 1924 Triangtois catalog, No. 14 features a distinctive castellated pediment on the second floor above the entrance. It opens in two sections to reveal six very large rooms and a hall with staircase. The attic is also accessible through the removable dormer. This house has windows on the sides and back as well as the front, providing natural light not often found in dollhouses. Each room contains a fireplace made by the Elgin firm. The house appears to be missing its original shutters ($1800-2000). 48" high x 55" wide x 20" deep. *House and photograph from the collection of Patty Cooper.*

The Tri-ang Princess Dollhouse was introduced some time in the early 1930s after Princess Elizabeth was given a child-sized Welsh cottage for her sixth birthday. It was produced for over two decades and was available in both front- and back-opening versions, with and without a garage. This model contains five rooms, including a bathroom, and also has a hall with stairway. The house opens in two parts. Three of the rooms contain Art Deco style fireplaces and the kitchen has a gas cooker, Welsh dresser, and sink. The house was wired for lighting, with battery compartments located under the removable chimneys not visible in the photograph. The rooms are quite large and easily accommodate furniture in 1" to the foot scale ($1200-1400). 28" high x 48" wide x 22" deep. *House and photograph from the collection of Patty Cooper.*

In 1939, Tri-ang introduced their flat roofed "Modern Dolls' Houses" in five different sizes. This No. 51 is the largest of the 3/4" scale series. As is often the case, the suntrap is missing from its roof ($700-800). 14" high x 26.5" wide x 10.75" deep. *House and photograph from the collection of Ray and Gail Carey.*

Inside the flat roofed house are four rooms and two hallways. Operable doors connect the rooms. This model features both a garage and a covered porch. The house was wired for electricity and still has its original Art Deco fireplaces. The stairway is missing. *House and photograph from the collection of Ray and Gail Carey.*

No. 61 of Tri-ang's popular series of small Tudor style houses features a covered entry porch and garage. The smallest, No. 60 in this series, had one gable and no garage. The largest, No. 64, had three gables and a garage. All of the houses were the same small 3/4" to the foot in scale, but varied in width according to the number of rooms. Tri-ang produced these houses from the early 1930s through the 1950s with only minor variations ($600-800). 16.5" tall x 19" wide x 10.75" deep. *House and photograph from the collection of Patty Cooper.*

No. 61 contains two rooms. The second story extends over the porch and garage on each side, but these spaces are somewhat inaccessible and difficult to decorate. The house contains furniture by Dol-Toi. *House and photograph from the collection of Patty Cooper.*

English Dollhouses and Furniture 43

No. 62 is the next larger model in the popular Tri-ang series. It opens in two sections and, unlike the smaller versions, this one contains a stairway ($700-900). 16.5" tall x 26.5" wide x 10.75" deep. *House and photograph from the collection of Ray and Gail Carey.*

The Tri-ang dollhouse known to collectors as the Stockbroker Tudor was advertised in their 1939 catalog as No. 93. The house was manufactured, with only minor changes, until the late 1950s. It opens in four sections to provide easy access to two bedrooms, a kitchen, and a living room, as well as the central hall and a bathroom over the garage. The house is properly scaled for the smallish 1" to the foot furniture manufactured by Tri-ang ($1500-1800). 24" high x 47" wide x 17" deep. *House and photograph from the collection of Patty Cooper.*

The Thistle version, produced by Lines Brothers in Canada, was back opening and slightly smaller. A Thistle advertisement from 1950 shows that, even in Canada, this model was numbered 60. One other dollhouse was shown in the Thistle advertisement, a slightly wider version, much like Tri-ang No. 61, with a single gable, covered porch, and garage. However, the Thistle version was numbered 21 and did not have the half-timbered design in the gable (Not seen often enough in U.S. to determine price). 15" high x 17" wide x 7" deep. *House and photograph from the collection of Mary Harris.*

The back of the house bears the label "Lines Bros. Canada/Thistle/Reg. Trade Mark/Made in Canada." Circa 1950. *House and photograph from the collection of Mary Harris.*

44 English Dollhouses and Furniture

By the 1950s, Tri-ang began producing dollhouses in contemporary styles. This model, with its metal front and non-opening windows, is a significant departure from earlier designs. It is listed as No. 50 in the 1950 catalog. The same style was manufactured until 1971, but the cardboard roof on this example indicates that it is one of the earlier ones. On later versions, the front door and roof were made of plastic ($350-500). 19" high x 16.25" wide x 9.5" deep. *House and photograph from the collection of Ray and Gail Carey.*

The metal front of No. 50 hinges open in one piece. The inside is quite plain, with four rooms and no interior doors or stairs. The ceilings are approximately 6" high making this house the appropriate scale for 3/4" to the foot furniture. *House and photograph from the collection of Ray and Gail Carey.*

Another house with a metal front was advertised as new in the 1964 Tri-ang catalog. Unlike earlier houses, this model has a sliding front which provides access to six rooms. The body of the house is made of hardboard ($350-500). 21" high x 26" wide x 11.5" deep. *House and photograph from the collection of George Mundorf.*

English Dollhouses and Furniture 45

A sample case opens to show the furniture produced by Eric Elgin of Enfield circa 1920 for Triangtois. The case shows the Jacobean dining room suite which included a combination fireplace and overmantel, along with a Jacobean bedroom suite (Too rare to determine price). *Private Collection.*

These pieces are typical of the kitchen furniture which often came in the 1" to the foot scale Tri-ang houses during the 1930s. The drawer on the Welsh dresser and the oven door on the gas cooker do not open ($20-25 each). Earlier houses often had a kitchen range built into the chimney surround. *Furniture and photograph from the collection of Patty Cooper.*

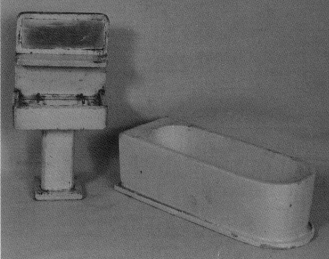

Several of the larger houses also contained bathrooms with fixtures glued in place ($20-25 each). In addition to the two pieces shown, a bathroom might have contained a toilet, medicine cabinet, and towel rack. *Furniture and photograph from the collection of Patty Cooper.*

The circa 1921 dining room pieces included the table ($40-45), sideboard ($45-50), and chairs ($20-25 each) shown here. The wardrobe from the Queen Anne bedroom suite is also shown ($45-50). *Private Collection.*

These dining room chairs, with Queen Anne legs, have the name "Elgin/Enfield" impressed under the seats ($20-25 each). Lines Brothers/ Triangtois sold furniture manufactured by the firm of Eric Elgin in Enfield until the Elgin factory closed in 1926. The furniture was made in a smallish 1" to the foot scale appropriate for most of the larger Tri-ang dollhouses. *Private Collection.*

The lid of a rare, boxed set of Triangtois furniture shows some of the other furniture produced in the early 1920s. This Jacobean bedroom suite is in the usual 1" to the foot scale (Too rare to determine price). *Private Collection.*

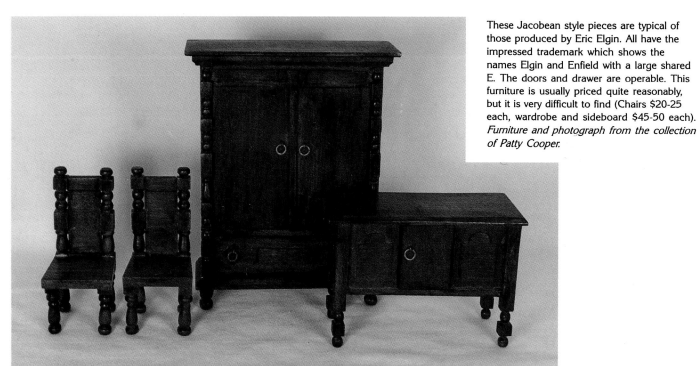

These Jacobean style pieces are typical of those produced by Eric Elgin. All have the impressed trademark which shows the names Elgin and Enfield with a large shared E. The doors and drawer are operable. This furniture is usually priced quite reasonably, but it is very difficult to find (Chairs $20-25 each, wardrobe and sideboard $45-50 each). *Furniture and photograph from the collection of Patty Cooper.*

English Dollhouses and Furniture 47

Lines Brothers continued to produce many of the Elgin designs and other similar pieces until World War II. These pieces are unmarked, but similar furniture is shown in the 1939 Tri-ang catalog. The blanket chest is listed as J54, denoting that it is one of the Jacobean style pieces ($45-50 each). *Furniture and photograph from the collection of Patty Cooper.*

The paper label under the seats of these dining room chairs reads "Period furniture. Made in England." The chairs, with their faux leather upholstery and "nail heads" suggested by dots of gold paint, are shown in the 1939 Tri-ang catalog. The Jacobean dining room set would have included four side chairs and two armchairs as well as a refectory table and court cupboard (Chairs $20-25 each). *Furniture and photograph from the collection of Patty Cooper.*

These Queen Anne style pieces, circa 1939, were upholstered in a small scale chintz pattern (Sofa $75-85, wing chair $65-75, ottoman $35, fire screen $50). *Furniture and photograph from the collection of Jim and Louise McCauley.*

The matching Queen Anne style bedroom included a bed ($85-100), writing armchair ($50-65), and dressing mirror ($45). *Furniture and photograph from the collection of Jim and Louise McCauley.*

BOX BACK DOLLHOUSES OFTEN KNOWN AS SILBER AND FLEMING TYPES

A large number of simple, boxlike dollhouses were manufactured in England from the mid-1800s through the early part of the twentieth century. Houses of this type have been found in catalogs from the firm of Silber and Fleming, a company involved in importing, wholesaling, and manufacturing. Although "Silber and Fleming" has become the most commonly used term to describe these houses, it seems unlikely that they were actually manufactured by that company. Instead, a variety of theories have been put forth suggesting that some or all of the houses may have been made by other known companies or in small workshops around London. Some of the houses which fall into the box back category were probably homemade. Given the large number of such houses; the great variations in size, style, and craftsmanship; and the number of decades they were produced, perhaps all of the theories have some substance. It should also be noted that Silber and Fleming was not the only company which sold houses of this type. Many of the early Lines houses had the same basic structure and several other firms advertised similar houses in their catalogs.

The houses in this category can be characterized by their simple, box-shaped backs and flat fronts. The facade was often higher than the main body of the house and almost all of the architectural detailing was found on the fronts of the houses, whose sides were usually plain. They were often painted or papered with an orange brick pattern on the second story and a cream-colored stone pattern on the first story. Sometimes the flat facades were relieved by applied trim around the doors and windows or balcony railings made of wood or metal. The houses were front opening and two or three stories high. The number of rooms varied from two to six, although it is possible that larger ones were produced.

As might be expected, the interiors were rather plain with boxy, windowless rooms. The proportions of the houses varied widely. Some had very shallow rooms with high ceilings which make them rather difficult for today's collectors to furnish. However, even the simplest houses usually had built-in fireplaces. Quite often they have been found with their original grates or even Evans and Cartwright kitchen ranges. Early houses were often painted inside, most typically with blue for the kitchens. Other houses contained wallpapers that were usually overscale, but subtle and delightfully typical of their time and place.

This tiny house appears to be missing its exterior paper, but its structure puts it firmly in the Silber and Fleming type category ($150-200). 13" high x 7.25" wide x 3.75" deep. *House from the collection of Lisa Kerr. Photograph by Bruce Kerr.*

This large Silber and Fleming type house is one of the more elaborate ones. It features two bay windows on the first story, Palladian windows on the second story, and three dormers on the roof. The house still has the original wallpapers and three faux marbleized fireplaces. There is a label on the back from Hamleys, the famous London toy store (Too rare to determine price). 42" high x 40.5" wide x 19" deep. *House from the collection of Anne B. Timpson. Photograph by Mary Kaliski.*

English Dollhouses and Furniture 49

This somewhat larger house displays a slightly different color scheme and pattern in the printed stones. Note how the proportions differ from the earlier houses. It is almost twice as tall, but only slightly wider and deeper ($650-800). 20.5" tall x 11" wide x 8.5" deep. *House from the collection of Dian Zillner. Photograph by Suzanne Silverthorn.*

The interior of the tiny house displays the overscale, but subtle wallpaper typical of Silber and Fleming type houses. *House from the collection of Lisa Kerr. Photograph by Bruce Kerr.*

The inside of this house displays a wallpaper pattern typical of Silber and Fleming type houses and two built-in fireplace surrounds. The inside of the opening front is also papered. *House from the collection of Dian Zillner. Photograph by Suzanne Silverthorn.*

This two-room house displays features typical of Silber and Fleming type houses. The elaborate printed pediments and pierced metal railing give the house a lot of character in a small, simple space. The house would best accommodate 1/2" to the foot scale furniture. 11" high x 10" wide x 5.5" deep ($550-700). *House and photograph from the collection of Mary Harris.*

50 English Dollhouses and Furniture

A very large box back house is shown to illustrate the variety of houses which fall into this category. This example may have been homemade, but since many of the Silber and Fleming type houses were believed to have been made as part of a cottage industry (rather than assembly line), the distinction may be somewhat nebulous. The windows are made of glass. Although the front door does not open, it is equipped with metal handles, a lock, and elaborate knocker (Not enough examples to determine price). 36" high x 34" wide x 12.5" deep. *House and photograph from the collection of Lois L. Freeman.*

Inside the large box back house are six rooms and a hall with stairway. Five of the rooms have fireplaces. The wallpaper appears to have been replaced circa 1940s. *House and photograph from the collection of Lois L. Freeman.*

AMERSHAM WORKS, LTD.

According to British researcher, Marion Osborne, the Amersham firm was established by Leon Rees to produce sporting equipment in the 1920s. By the early 1930s, the company began manufacturing dollhouses and toy garages, forts, and airports. The houses were made of wood with metal windows. They ranged in size from a simple one-room cottage to a 56" long house with six rooms and a garage. A variety of styles were produced, including a "sun house" which was very similar to the Tri-ang flat roofed houses. Most of the Amersham houses were Tudor in style, which makes them easily confused with those produced by Tri-ang. Luckily, many Amersham dollhouses still retain their original label proudly displayed, front and center, on the base.

The windows are the features which best distinguish these houses from those made by other British companies during the same period. Many of the Amersham houses had square paned windows which opened on a center pivot rather than being hinged. Other houses had lattice style windows in which the lower sections opened on hinges at the side but the transoms at the top were fixed in place. This makes them easily distinguishable from similar Tri-ang or Romside windows, which are hinged at the sides and do not contain the transoms.

This circa 1930s Amersham house displays the square paned windows which pivot in the center ($450-600). 14.75" high x 14.25" wide x 10.25" deep. *House and photograph from the collection of Sharon and Kenny Bernard.*

English Dollhouses and Furniture 51

The two room interior is enlivened by the colorful wallpaper and electric lights. *House and photograph from the collection of Sharon and Kenny Bernard.*

A close-up shows the nursery wallpaper features Golliwogs and other toys. *House and photograph from the collection of Sharon and Kenny Bernard.*

Fortunately for collectors, many of Amersham houses bear a prominently displayed metal label. *House and photograph from the collection of Sharon and Kenny Bernard.*

Some Amershams had rather complex roofs with multiple gables forming several distinct peaks from front to back. Often, but not always, Amersham houses had overlapping strips of wood on the roofs to give the impression of overscale shingles. Other Amersham roofs had an incised pattern intended to represent tile.

Amersham houses have been found with wallpapers containing a nursery motif of golliwogs or children playing. The printed floor papers had a parquet pattern. Some of the fireplaces were almost whimsical, very tall with exaggerated curved lines, reminiscent of the Art Nouveau style. Many of the houses included garages, sun porches, and bay windows. They were equipped with electric lights and, according to one advertisement, electric fireplaces. It is not known when the company went out of business, but there is no record of Amersham dollhouses being sold after the 1950s.

Another house from the 1930s has a roof which is typical of Amershams with its front to back gables and overlapping wood strips. The exterior is decorated with sponged vines and plants ($450-550). 18" high x 17.75" wide x 12" deep. *House from the collection of Lisa Kerr. Photograph by Bruce Kerr.*

52 English Dollhouses and Furniture

There are four rooms, including the garage, inside the sponge decorated Amersham. Some of the paper has been replaced ($450-550). *House from the collection of Lisa Kerr. Photograph by Bruce Kerr.*

A similar Amersham has four rooms with a garage extension. Although the house has been repainted, it is still uniquely Amersham ($350-500). *House from the collection of Sharon and Kenny Bernard. Photograph by Gail Carey.*

HOBBIES

Hobbies Weekly was a popular magazine published in Dereham, Norfolk, from 1895 to 1970. They also published an annual supplement called the *Hobbies Handbook*, which often contained plans for the construction of dollhouses and dollhouse furniture. In addition to the patterns, Hobbies provided a source for papers which could be used on the roofs and exteriors. It is difficult to know how many English dollhouses from early in this century may have been made from *Hobbies* plans or what these early houses looked like.

A 1935 issue of *Hobbies* showed a plan for a 1" scale dollhouse 24" high and over 30" wide, with four rooms, a hallway, and landing. This house was open backed and Tudor in style. Designs for twenty-six pieces of "modern style" furniture were included in the package along with the necessary wood, glass, mirrors, hinges, and papers. The option of purchasing either the wood or the fittings separately was also offered.

A two-room Hobbies dollhouse features a garage and walled garden area ($350-450). 20" high x 22" wide x 16" deep. *House and photograph from the collection of Ruth Petros.*

English Dollhouses and Furniture 53

During the 1930s, they began offering a range of metal doors, windows, stairs, and other parts to be used in the construction of dollhouses, in addition to the papers. The metal parts were made by Romside Manufacturing Ltd. This company produced metal doors and windows for dollhouses from the 1930s through the early 1970s. The Romside windows are easily identified by their side hinges and butterfly knobs. The mullions were diamond shaped until the 1960s, when they began to produce windows with square panes. The Romside doors are even more distinctive, with two panels below a letter slot and latticed window. However, Hobbies was not the only company which used Romside products. The components could be purchased in hobby shops and were also used by other British dollhouse manufacturers including Tudor Toys. The Romside parts gave the Hobbies houses the appearance of having been commercially made, which is why the houses are being included in this book.

Surprisingly, quite a number of Hobbies houses have been found, even in America. Most of these were probably made in the 1950s and early 1960s. Although most appear to have been constructed pretty much "by the book," the fact that these were plans rather than kits allowed a great latitude in interpretation. This creates a problem for the researcher in defining a Hobbies house. To what degree must the maker have adhered to the pattern for the house to be called a Hobbies? Is it a Hobbies house if one of their patterns was followed, but none of their papers or components used? And, of course, it would have been possible to use all the Hobbies type components, as produced by Romside, on a dollhouse never dreamed of by Hobbies. In addition to all this, the quality of the finished house varied according to the skill of the maker.

The original *Hobbies* ceased publication in the 1960s. In 1992, the Stroulger family, which had been associated with *Hobbies* since the 1930s, began publishing the magazine once again.

A circa 1950s Hobbies house, with printed brick paper, opens in two sections ($600). 23" high x 24" wide x 17" deep. *House from the collection of Dian Zillner. Photograph by Suzanne Silverthorn.*

Inside the circa 1950s Hobbies house there are four rooms plus two hallways which contain the staircase. A bathroom has been tucked into the stair landing. This house contains original floor papers, some original wallpapers, and four built-in Romside fireplaces. *House from the collection of Dian Zillner. Photograph by Suzanne Silverthorn.*

The front of this four-room Hobbies house, circa 1960s, opens in two sections. The Romside front door, windows, and garage doors, which were offered by the magazine, can be clearly seen. This example does not have exterior papers ($300-450). 18.5" high x 29.25" wide x 10.5" deep. House from the collection of Dian Zillner. *Photograph by Suzanne Silverthorn.*

The later, square paned Romside windows are used on this circa 1960s house which is believed to be a Hobbies. This example is front opening and contains four rooms. The ceiling are 5.25" high, making it appropriate for small 3/4" to the foot scale furniture ($400-500). 16.5" high x 23.5" wide x 12" deep. *House and photograph from the collection of Ray and Gail Carey.*

TUDOR TOY COMPANY AND OTHERS

Until recently, little was known about the Tudor Toy Company dollhouses and often they were attributed to Tri-ang. In the early 1990s, Marion Osborne identified these houses and began sharing her research through articles published in English dollhouse magazines. The Tudor Toy Company began manufacturing dollhouses in 1946 and sold them under the brand name GeeBee until 1978. At that time, the company was taken over by Humbrol, who continued to produce the houses for several more years. The early dollhouses were made of wood and had metal windows. In the 1960s, the company began to make greater use of fiberboard in parts of the houses and by the 1970s, some of the windows were plastic. Originally, the fronts of the houses opened on hinges. Later houses incorporated other means of access, including dowels used as pivots, sliding doors, and open backs. Many of the Tudor Toys houses have a painted vine decorating the fronts.

As might be expected, given the name of the company, the houses were often Tudor in style, but the company produced some houses which might best be described as "Swiss chalets" and some that were contemporary in design. Among the contemporary models was a "California Bungalow," a one-story house complete with swimming pool. In the 1970s, the company produced dollhouses of lithographed hardboard.

It is very confusing to try to identify a Tudor Toys house. The company used metal windows, manufactured by Romside, which were hinged at the sides and usually had diamond shaped mullions. This feature helps to distinguish them from Tri-angs, with their square paned windows, and Amershams, whose windows opened on a center pivot. Unfortunately, for purposes of identification, Romside windows were included in the Hobbies dollhouse kits and also used by at least two other British dollhouse manufacturers, including Pennine Products and the even more obscure Conway Valley Series. To further complicate matters, some of the Tudor Toys houses produced in the 1960s had square paned windows. There are a few other characteristics which may provide clues to identifying Tudor Toys houses. The early houses sometimes had strips of wood which extended above the base to form a low "wall" on each side of the house. The interiors were usually painted rather than papered like Tri-angs or Amershams. Many of the houses manufactured in the 1960s featured retractable awnings. A telltale slit found over the windows of an unknown house would be a good indication that it was made by Tudor Toys. And, of course, sometimes they still have their labels!

The maker of this two-room house is unknown, but the diamond paned windows and painted interior suggest that it may be a product of the Tudor Toy Company ($450-500). 15" high x 13" wide x 7.5" deep. *House from the collection of Zelma Fink. Photograph by Kathy Garner.*

Similar houses were shown in Tudor Toys advertisements from the early 1950s. The awnings, windows, and side "walls" of the garden are distinctive Tudor Toys features ($450-600). 16.5" high x 23.25" wide x 8.5" deep. *House and photograph from the collection of Betty Nichols.*

This Tudor Toys house opens in two sections. The plain, un-papered walls are typical of Tudor Toys houses. *House and photograph from the collection of Betty Nichols.*

An example from the mid-1960s features the label "Tudor Lodge" on its front ($350-500). 24.5" high x 19" wide x 13.5" deep. *House from the collection of Ann Hurley. Photograph by Karen Evans.*

A smaller "Tudor Cottage" was advertised by Tudor Toys from the mid-1960s through the mid-1970s ($250-400). 19" high x 16.5" wide x 11" deep. *House and photograph from the collection of Ray and Gail Carey.*

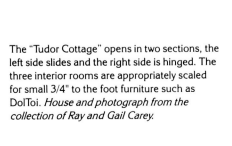

The "Tudor Cottage" opens in two sections, the left side slides and the right side is hinged. The three interior rooms are appropriately scaled for small 3/4" to the foot furniture such as DolToi. *House and photograph from the collection of Ray and Gail Carey.*

Tudor Toys advertised this four-room house in 1966. It opened from the front in two sections and contained a stairway ($225-250). 16" high x 23.5" wide x 11.5" deep. *Courtesy of Left Bank Antiques, Anacortes, Washington. Photograph by Gail Carey.*

This small, marked Tudor Toys house was advertised in 1966. It has the later Romside, square-paned windows. The front opens in one section and there are two, unpapered rooms inside. The awning which originally shaded the first floor window is missing ($165-200). 16" high x 14" wide x 11" deep. *Courtesy of Left Bank Antiques, Anacortes, Washington. Photograph by Gail Carey.*

If not for its metal label, which reads "Pennine. Made in England," this house might easily be confused with a Tudor Toys or Tri-ang. The Pennine company made dollhouses circa mid-1950s through early 1960s. Inside the Pennine house are two unpapered rooms. The upstairs room contains wings on each side which are difficult to access ($350-500). *House and photograph from the collection of Mary Harris.*

This four-room house displays some features characteristic of Tudor Toys houses, such as the metal, diamond paned windows and painted interior. However, its unusual roof design suggests that it may have been made by another, lesser known British company, perhaps Conway Valley or Pennine. The interior partitions of the house are movable ($350-500). 18" high x 19" wide x 13" deep. *House from the collection of Linda Boltrek. Photograph by Lois Freeman.*

A small, two-room cottage bears the Pennine label above the second-story Romside windows ($150-200). *Courtesy of Left Bank Antiques, Anacortes, Washington. Photograph by Gail Carey.*

A. BARTON AND CO. (TOYS) LTD.

The furniture made by the English firm of A. Barton and Co. (Toys) Ltd. is relatively easy for the collector to find because it was manufactured for many years, in great quantity, and exported to the United States. British researcher, Marion Osborne has provided the most complete source of information about the company through her book *Barton's Model Homes*. In it, she describes how Albert Barton established a factory on May 8, 1945, the day peace was declared in Europe. During the war, he had been involved, on a smaller scale, in making toys under the trade name Peko.

Barton's utilitarian furniture styles exemplify those popular with the middle class during the post-war period, from the company's establishment in 1945 until 1984. Throughout the almost four decades that the Barton Company was in business, their furniture underwent relatively minor changes. Scale remained a fairly consistent 3/4" to the foot, but the materials changed and styles were somewhat updated. The furniture was sturdily built of plywood, with some wood parts and cardboard backs. Most of the drawers and doors opened, a feature not found in most of the dollhouse furniture being made in America at the same time and in the same scale.

It is extremely difficult to date any piece of Barton furniture precisely, as most were made for many years, with new styles overlapping older ones.

Generally, a large number of pieces were manufactured during the 1940s and 1950s, with minimal stylistic changes. Most of the earlier furniture had a medium stain (called "mahogany" by the company); later pieces were quite light ("beech") and more Danish Modern in style. Kitchen appliances were enameled in off-white and red or green. Originally, strips of wood were used as handles. Some of these were replaced with round knobs by the 1950s. The bathroom fixtures made throughout most of this period were plaster. The early logo was a diamond shape with the letters "A" and "B" fitted into the top half. Furniture was packaged in a box, printed with the diamond logo and the words "Model Home. Scale Model Furniture." Individual pieces were also sold.

In addition to the products they manufactured, Barton sold furniture and accessories made by other small firms and cottage industries. Most notable among these were the metal items made by Taylor and Barrett, and after the war, plastic accessories made by A. Barrett & Sons (Toy) Ltd. and F.G. Taylor and Sons Ltd.; the Grecon dolls made by Grete Cohn (see the Dollhouse Doll chapter); wire and plastic items produced by Mr. Rickard; and plaster items by Kaybot Novelties. Many of these were imported into the United States and sold by other distributors.

One of the most popular sets produced by Barton, and most confusing to collectors, was their Tudor line. This heavy, very dark stained furniture, in Tudor and Jacobean period styles, resembled furniture manufactured earlier by Elgin and Tri-ang. However, unlike that produced by the other companies, the Barton furniture was 3/4" to the foot in scale. This furniture is often thought to be much older than it really is. It was made by Barton from 1948 through the late 1970s. Its small scale and period style make it highly sought by collectors furnishing small antique dollhouses. Considering the length of time it was produced, the variety of pieces was rather limited. The more common pieces include a trestle table and chairs, monk's bench, blanket chest, court cupboard, and four poster bed, which was available both with draperies and without. More difficult to find are the sideboard, Welsh dresser, demi-lune table, grandfather clock, tallboy, dressing table, corner cupboard, fireplace, corner seat, and settle. Some of the Tudor items were exported to America with an unstained, natural finish.

In the mid-1950s, the Barton company introduced a very modern, open backed dollhouse, mostly available in kit form. The house had an unusual, split-level room with a stairway. Under this room, there was a drawer which could be removed in order to use the space as a garage. A very similar, ready made model was available in the 1960s along with a four-room bungalow. By the 1970s, the storage drawer was eliminated.

By the mid-1960s, a much more limited range of items was available. The logo became a very rounded lower case "a" connected to a capital "B" and the boxes were changed to bubble-style packages labeled "Model Home series." Although most of the pieces were still made of plywood, there were many more plastic parts and accessories. The nursery pieces were a creamy yellow and the kitchen furniture a pure white with red plastic doors. The bathroom set was plastic. In addition to the sets illustrated, the back of the box showed a stove/dishwasher set and a dining room set.

The Caroline's Home dollhouse was introduced in 1975 along with the new line of Caroline's Home furniture. Many of the first Caroline's Home

These pieces are typical of those produced by Barton from the 1940s through the early 1960s. The early pieces had a "mahogany" finish (Chairs $10-12, other pieces $15-18 each). *Furniture from the collection of Carol Miller. Photograph by Jeff Jackson.*

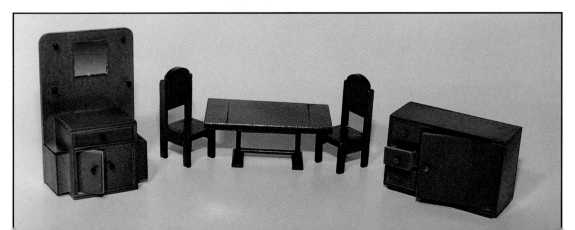

58 English Dollhouses and Furniture

pieces were made of plywood. These included the dining room, which was basically the same as the one shown on the back of the "Model Home" bubble packages. Both the bathroom and kitchen pieces, while retaining a similar design, were produced in plastic. By the 1980s, only the bases of the bedroom and living room furniture were plywood. Everything else was plastic, including the dining room in a surprising Regency style, which contrasted sharply with the modernistic design found in all the other sets. The furniture was packaged in boxes with cellophane windows. The boxes proclaimed that "Caroline's Home Interior Fashions have detailing just like full-size furniture," and amazingly, even the plastic pieces were highly detailed with opening drawers and doors.

In 1984, the Barton company was purchased by Lundby. The Caroline's Home furniture continued to be distributed by the new owners for nearly another decade.

Scaled 3/4" to the foot, these bedroom pieces are typical of those produced by Barton from the 1940s through the early 1960s. Note the wardrobe, which continued with few changes, throughout most of Barton's production ($15-18 each). *Furniture from the collection of Carol Miller. Photograph by Jeff Jackson.*

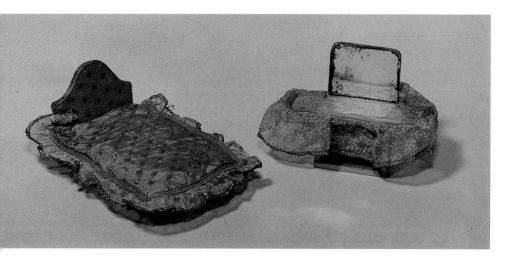

The Barton firm produced various "trimmed" beds and dressing tables throughout most of its history. This set is believed to be from the 1950s and was acquired along with other Barton pieces ($15-18 each). *Furniture from the collection of Carol Miller. Photograph by Jeff Jackson.*

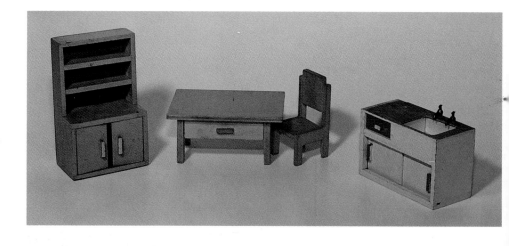

Typical of kitchen pieces produced by Barton in the 1940s through 1950s, these show two of the main color combinations (Chairs $10-12, other pieces $15-18 each). *Furniture from the collection of Carol Miller. Photograph by Jeff Jackson.*

English Dollhouses and Furniture 59

Similar nursery furniture was produced for many years including this "drop-side cot" and highchair ($12-15 each). Other nursery pieces included a cradle, chest of drawers, and rocking horse. *Furniture from the collection of Carol Miller. Photograph by Jeff Jackson.*

The Barton bathroom pieces remained much the same over several decades with a long narrow bathtub, low toilet, and top heavy sink. The colors changed over the years. This plaster set is probably from the 1960s (Set $35). *Furniture from the collection of Carol Miller. Photograph by Jeff Jackson.*

The plaster fireplace on the left dates from the mid-1960s ($20-25). The "tiled" wood fireplace on the right is probably from the early 1950s ($12-15). After that time, the printed flames disappeared (the fender on this piece has been replaced). The stove in the middle has been attributed to Barton, but not verified in catalog photos ($15-18). *Furniture and photograph from the collection of Patty Cooper.*

This set of cleaning tools from the late 1950s came in a red box with a cellophane window, typical of Barton packaging from this time ($30-35). *Boxed set from the collection of Dian Zillner. Photograph by Suzanne Silverthorn.*

Barton also sold items made by a man named Mr. Rickard. The metal washing machine has been attributed to him ($35-40). Other items were sold by Barton but may have been manufactured by some of their other suppliers ($15-25 each). *Items from the collection of Carol Miller. Photograph by Jeff Jackson.*

The early metal fireplaces and "gas" stoves were produced for Barton by Taylor and Barrett, circa 1940s. Generally, pieces made before the war were marked "T&B" while those made after the war were marked "B&S" for A.Barrett and Sons ($30-35 each). The stove on the left, with red plastic door and plastic pots, is from the 1960s ($18-20). *Furniture and photograph from the collection of Patty Cooper.*

Two bubble-packaged sets, circa 1960s, contain some of the pieces which are more difficult to find in the United States, including the Welsh dresser, fireplace, and corner cupboard (Sets in original packages $95-110). *Furniture and photograph from the collection of Ruth Petros.*

Barton sold a wide variety of accessories. The cutlery set was sold for nearly four decades. The red metal tray on the left was probably manufactured by Barrett until 1970 using prewar molds. The toaster and aquarium were also produced by Barrett and sold as novelty items (Items in original packaging $20-25, other pieces $8-12 each). *Accessories and photograph from the collection of Patty Cooper.*

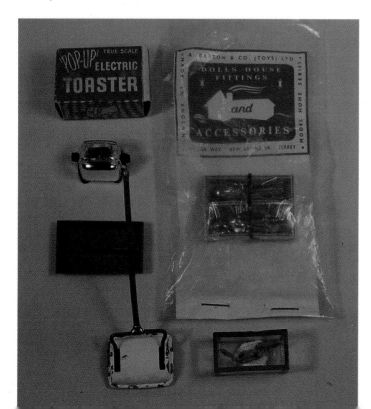

The Barton Tudor line was produced from 1948 through the late 1970s. This photo shows some of the more commonly found pieces. The bed was sold with and without draperies. These may not be original (Chairs $10-12, bed $20-25, other pieces $18-22). *Furniture and photograph from the collection of Patty Cooper.*

English Dollhouses and Furniture 61

This circa 1966 bedroom set is most notable for its spindly (almost like toothpicks) legs. These legs were used briefly on a few other pieces during this period ($10-12 each). *Furniture and photograph from the collection of Patty Cooper.*

The "radiogram" was introduced in the late 1960s and was still being used in the early Caroline's Home line. The case was made of wood with a plastic turntable. Accessories included a floor lamp, flower pot, and mirror (Packaged set $40-50). *Furniture and photograph from the collection of Patty Cooper.*

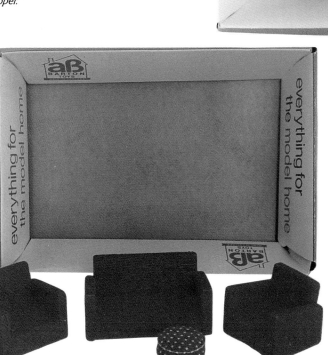

The three-piece living room suite from the early 1970s is flocked in red. The polka dotted ottoman was continued into the early Caroline's Home sets (Set in original package $50-60). *Furniture and photograph from the collection of Patty Cooper.*

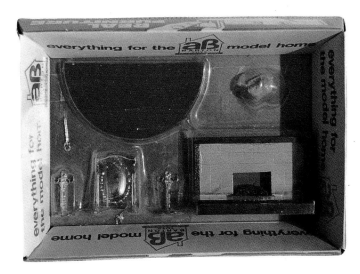

By the late 1960s the fireplace no longer showed a fire. Other accessories in this set included a rug, fire screen, and tools (Packaged set $40-50). *Furniture and photograph from the collection of Ray and Gail Carey.*

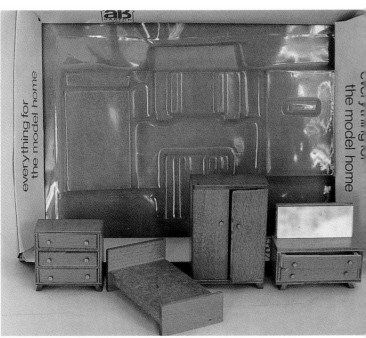

The early 1970s bedroom was made of wood, plywood, and chip board. The doors and drawers opened with round brass knobs. The bed originally had a foam rubber mattress which has deteriorated (Packaged set $50-60). *Furniture and photograph from the collection of Patty Cooper.*

The early 1970s sink had a plastic top on a wood base. The toaster and plastic food were included with the boxed set. A separate set (not shown) was required to complete the kitchen (Packaged set $50-60). *Furniture and photograph from the collection of Patty Cooper.*

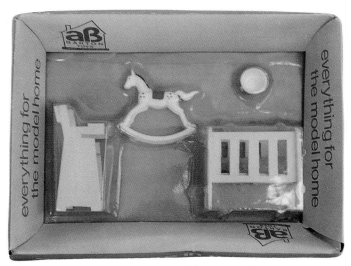

The early 1970s nursery set included a "drop-side cot," high chair, plastic rocking horse, and potty (Packaged set $40-50). *Furniture and photograph from the collection of Patty Cooper.*

English Dollhouses and Furniture 63

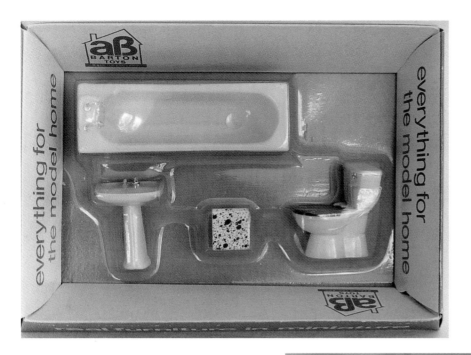

The early 1970s bathroom set was molded in pink plastic. The sides of the bathtub and the hinged toilet lid are black. The flush lever actually moves a ball cock inside the tank, although this is only noticeable if the piece is held up to the light. Only the bath stool is made of wood, with a faux cork top of speckled paper (Packaged set $40-50). *Furniture and photograph from the collection of Patty Cooper.*

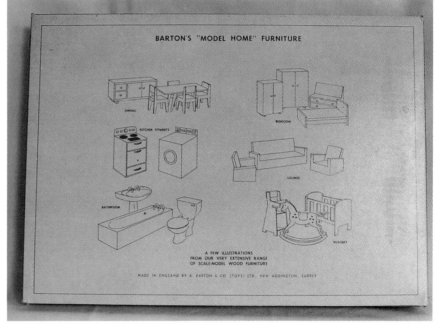

The back of the early 1970s bedroom box illustrates the pieces missing from this collection. These include a stove, washing machine, and dining set. This dining room furniture was also used with the early Caroline's Home line. *Furniture and photograph from the collection of Patty Cooper.*

The early Caroline's Home living room pieces, circa 1975, were made mostly of wood. The "radiogram" was replaced by a television in the later Caroline's Home sets. The piano is an especially desirable piece (Piano $20, other pieces $10-14 each). *Furniture from the collection of Dian Zillner. Photograph by Suzanne Silverthorn.*

The Caroline's Home dollhouse, introduced in 1975, was made of plastic and cardboard. Access was through the open front and right side. The house was equipped with lamps that lighted the house. A "super deluxe" version was added in the early 1980s ($125-150). 16" high x 27" wide x 11" deep. *House from the collection of Dian Zillner. Photograph by Suzanne Silverthorn.*

Barely visible in the photograph is the bedroom wall decorated with a very large horse and rooster. Interior doorways are open with rounded tops. The house contained six rooms. *House from the collection of Dian Zillner. Photograph by Suzanne Silverthorn.*

The first Caroline's Home bedroom contained several pieces which were similar to earlier lines. The wardrobe remained virtually unchanged from the one found in the 1940s sets. The same dressing table design was used in sets from the early 1970s ($10-14 each). *Furniture from the collection of Dian Zillner. Photograph by Suzanne Silverthorn.*

Barton continued to produce a version of its "trimmed bedroom" in the first Caroline's Home line, circa 1975 ($10-14 each) *Furniture from the collection of Dian Zillner. Photograph by Suzanne Silverthorn.*

The first Caroline's Home dining room (circa 1975) was very much like that from earlier Barton sets (Chairs $10 each, table and buffet $12-14 each). The dolls ($18-20 each) appear to be the early Caroline's Home dolls which were later replaced by Lundby dolls. *Furniture from the collection of Dian Zillner. Photograph by Suzanne Silverthorn.*

The "electric" stove and refrigerator retained the same design found in pieces from the early 1960s, but the Caroline's Home pedestal table and chairs were a radical departure from previous Barton kitchen pieces. All of these pieces were made of plastic (Chairs $8, other pieces $10-14). *Furniture from the collection of Dian Zillner. Photograph by Suzanne Silverthorn.*

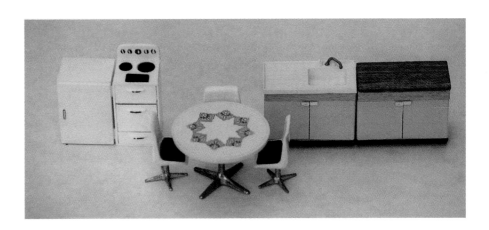

The circa 1985 version of the Caroline's Home kitchen was all plastic and, for the first time, included overhead cabinets, very much like Lundby's. This boxed set contained only a cabinet, sink, and washing machine—no stove or refrigerator. Unlike earlier kitchen pieces, the doors on this set are hinged rather than sliding (Packaged set $25-30). *Furniture and photograph from the collection of Patty Cooper.*

66 English Dollhouses and Furniture

The later Caroline's Home living room contained a modernized fireplace with plastic grate and three upholstered pieces with oversized fringe (Packaged set $25-30). *Furniture and photograph from the collection of Patty Cooper.*

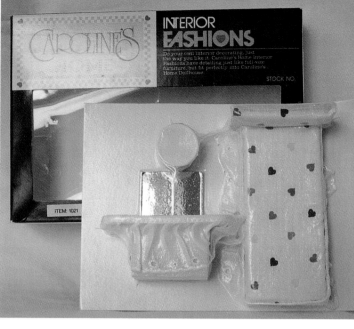

The Caroline's Home bedroom introduced in 1985 was made of chunky blocks of wood covered in fabric (Packaged set $25-30). *Furniture and photograph from the collection of Patty Cooper.*

A different Caroline's Home bedroom set, in the same packaging, was probably made a year or two later. It is made of wood and the vanity has an opening drawer (Packaged set $25-30). *Furniture and photograph from the collection of Ray and Gail Carey.*

The most drastic change in the later Caroline's Home furniture was in the design of this all plastic Regency style dining room set which hardly seemed to belong in the same line as the chunky bedroom (Packaged set $25-30). *Furniture and photograph from the collection of Patty Cooper.*

DOL-TOI PRODUCTS (STAMFORD) LTD.

The English firm of Dol-Toi made dollhouse furniture and accessories in a scale which ranged from 1/2" to one inch equals a foot. Easily confused with Barton, Dol-Toi furniture was slightly smaller and somewhat less detailed. The Dol-Toi pieces were made of beech wood and the doors and drawers were operable. The company began making furniture in 1945 and exported it to the United States. The furniture was packaged in boxed sets or individually shrink-wrapped. Originally, the furniture was marked with an ink stamp, later with a round paper label bearing the words "Dol-Toi. Made in England." The early lines of furniture were varnished dark brown, but by the 1960s, the furniture had a very light, natural finish. The kitchen and bathroom pieces were white or off-white and nursery pieces were enameled in blue or pink.

Dol-Toi produced a wide range of styles for the dollhouse, both contemporary and traditional. The 1964-65 catalog showed ten different bedrooms, ranging from the very traditional "draped" set with canopy bed through the "contemporary" set with bunk beds. Seven different styles of dining rooms and three different designs of sitting rooms coordinated with the bedroom sets. There were three kitchen sets, one containing a gas-type stove and wringer washing machine, while another had an Aga-type solid fuel cooker. The variety of styles available in one year should warn the collector against presuming to date a piece of Dol-Toi based on style alone.

Dol-Toi also made a line of accessories in both 3/4" and 1" scales. The accessory packages listed a wide variety of items including cushions, six different rugs, bedding, nine sets of food, various pots and pans, pets, telephone, bird cage, ironing board, cleaning tools, and many other kitchen items.

The Dol-Toi line included flexible dollhouse dolls, made of yarn covered wire, with adults approximately four inches tall. The early dolls had faces made of cloth covered buttons. The features were stamped onto the fabric, with the eyes, which resemble sideways commas, always looking to the doll's right. Their small scale, yarn hair, and metal feet make them easily confused with Grecon dolls, also produced in England. When the supplier was unable to keep up with the demand, the company began making dolls with molded plastic heads which seem disproportionately small compared to their feet.

The company was sold in the mid-1970s and by 1978 it was no longer producing dollhouse furniture.

The early Dol-Toi living room pieces were dark stained with strips of wood for pulls on the operable doors and drawers ($12-15 each). Sofas and chairs were flocked ($10-12 each). The fireplace is made of plaster ($12-15). *Furniture and photograph from the collection of Patty Cooper*

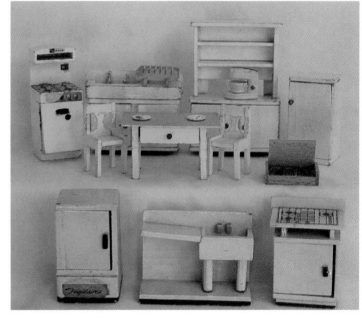

A wide variety of kitchen styles were available over the years. Printed paper on the tops of the stoves created fairly detailed cooking surfaces (Chair $8-10, other pieces $12-15 each). *Furniture and photograph from the collection of Patty Cooper.*

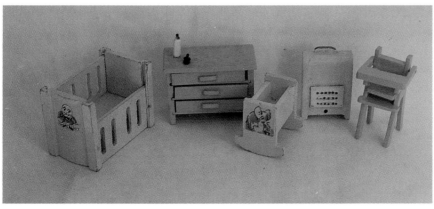

The nursery pieces were enameled in pink or blue with paper decals showing animals or nursery rhyme figures ($12-15 each). *Furniture and photograph from the collection of Patty Cooper.*

68 English Dollhouses and Furniture

The Dol-Toi pianos had paper keyboards with opening covers ($20). The televisions came in several different styles ($12-15). None showed a picture on the screen. *Furniture and photograph from the collection of Patty Cooper.*

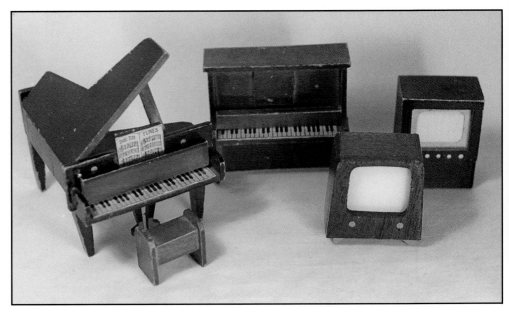

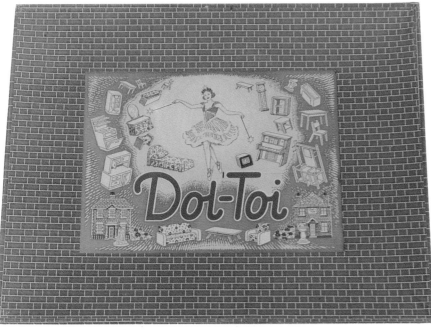

The furniture was available in boxed sets. Other pieces in the line are shown on the cover (Boxed set $100). *Furniture and photograph from the collection of Patty Cooper.*

This bedroom set is typical of Dol-Toi's "contemporary" line, modern in style and light in color. This set includes a plaster fireplace and metal telephone, probably made by Barrett & Sons. *Furniture and photograph from the collection of Patty Cooper.*

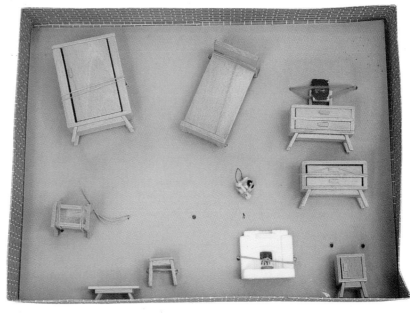

English Dollhouses and Furniture 69

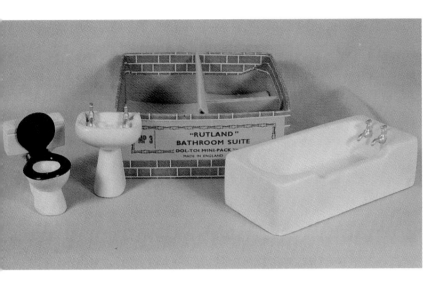

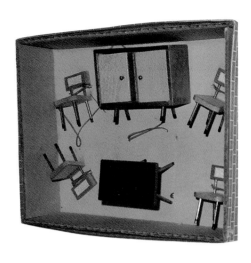

The "Rutland" bathroom set was shown in the 1964-65 Dol-Toi catalog. It was made of plaster and the toilet lid and seat could be lifted (Boxed set $45). *Furniture and photograph from the collection of Patty Cooper.*

The "Continental Dining Suite" was also shown in the 1964-65 Dol-Toi catalog. The pieces were made of wood and the doors of the buffet opened (Boxed set $65). *Furniture and photograph from the collection of Ruth Petros.*

By the 1960s, individual pieces were available, shrink-wrapped, with a round paper label ($18-20 each in original wrapping). *Furniture and photograph from the collection of Patty Cooper.*

These pieces are from the "Draped Bedroom Suite" shown in the 1964-65 catalog ($12-15 each). The canopied bed is missing from this set. *Furniture and photograph from the collection of Patty Cooper.*

70 English Dollhouses and Furniture

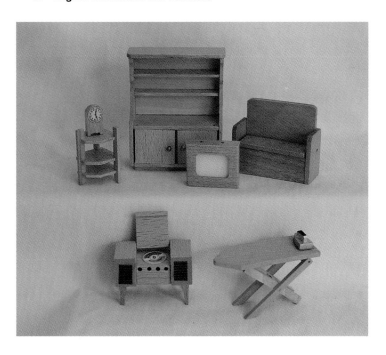

Later additions to the line included pieces with a very light finish. Round brass knobs replaced the strips of wood previously used for pulls (Accessories $5, other pieces $10-15 each). *Furniture and photograph from the collection of Patty Cooper.*

The metal wringer washer was probably made for Dol-Toi by Barrett & Sons ($20-25). *Furniture and photograph from the collection of Gail Carey.*

Dol-Toi kitchen accessories included boxed sets of metal saucepans with plaster food (Boxed set $25, individual pieces $3-5). *Accessories and photograph from the collection of Patty Cooper.*

Dol-Toi's full line of accessories even included linens ($10 in original package). *Accessory and photograph from the collection of Patty Cooper.*

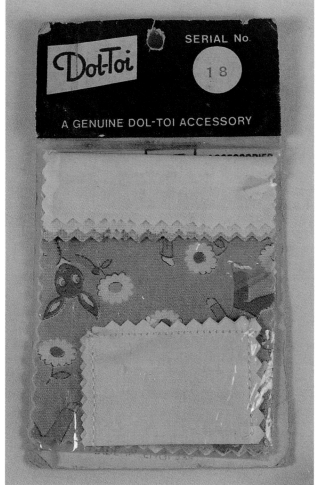

English Dollhouses and Furniture 71

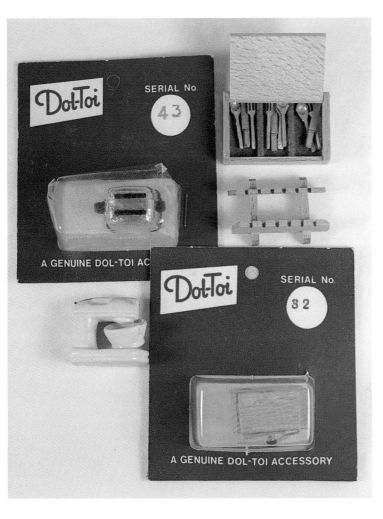

Kitchen accessories were quite detailed, down to the silverware box lined in felt and containing tiny, individual utensils (Silverware box with hinged lid $12, mixer $6, dish rack $4, pieces in original bubble packs $8-10 each). The toaster was made by Barrett & Sons and is nearly identical to the one sold by Bartons ($12). *Accessories and photograph from the collection of Patty Cooper.*

Dol-Toi dolls were made of flexible wire wrapped with yarn. Adult dolls were approximately four inches tall. The child on the left is an earlier doll whose head is made of a rubber button covered with fabric. The features are stamped on. The other four dolls have heads molded in a plastic resin material. All of the dolls have heavy metal feet which allow them to stand unsupported ($18-20 each). *Dolls and photograph from the collection of Patty Cooper.*

OTHER BRITISH DOLLHOUSE FURNITURE

Several other British companies are known to have made dollhouse furniture in metal, wood, or plastic. In addition, collectors occasionally encounter pieces of furniture, made by companies as yet unknown, which are simply marked "England" or are found in English dollhouses or appear to be British in style.

Airfix
Airfix was an early producer of plastic dollhouse furniture. The company advertised a set of furniture in 1947 and made bedroom and living room furniture by 1954. Airfix also manufactured plastic dolls and an ice cream tricycle which could be used with dollhouses.

Airfix made plastic furniture in a somewhat inconsistent 3/4" to the foot scale circa 1954. The table is 4.5" long but only 1.5" tall (Chairs $5, table and buffet $8 each). *Furniture and photograph from the collection of Ray and Gail Carey.*

Bex

British Xylonite Company, Ltd. of London was an early manufacturer of Bakelite dollhouse furniture under the tradename Bex. Margaret Towner has identified the furniture as having been manufactured circa 1937. The fluidity of Bakelite allowed the furniture to be produced in an Art Deco style which must have seemed outrageously modern for its time. Sets were made in red, green, blue, and brown. The pieces shown are marked "A BEX moulding/Made in England."

Charbens & Co. Ltd.

The Charbens Company, which was located in London, began making small scale, cast metal furniture in the 1930s. The furniture was approximately 1/2" to the foot in scale and fairly detailed. The parlor furniture shown here has a realistic upholstery pattern cast into the metal. The company continued operating into the 1970s, but it is not known how long they produced dollhouse furniture.

Crescent Toy Co. Ltd.

The Crescent Toy Company, Ltd. of London and Wales manufactured small-scale, cast metal dollhouse furniture in the 1950s, usually marked "D.C.M.T." or "Crescent." The company also produced metal soldiers and other items.

Evans & Cartwright

The Evans & Cartwright tinplate dollhouse furniture was produced in Wolverhampton circa 1820-1880. The furniture was made in two scales, the largest approximately 1.25" to the foot. It is highly collectible and usually quite expensive. The name "Evans & Cartwright" is sometimes found, in very small letters, cast into the metal. This furniture has sometimes been referred to as "Orley." Some wonderful examples can be seen in the book *The Vivien Greene Dolls' House Collection*. The furniture was made in a variety of styles and finishes. The most commonly found has a faux mahogany brownish-orange finish. The company also made distinctive fireplaces and kitchen ranges which occasionally are found in English dollhouses from the period.

Fairylite

Fairylite was the tradename used by Graham Bros., an importing firm in London. Founded in the 19th century, the company was in business until the 1950s. They sold a variety of dollhouse furniture both made in England and imported from Japan. One of their best known products was the cast metal bathroom set shown here. It was manufactured in approximately 3/4" to the foot scale in green or cream. This set has been found marked "Fairylite" and also with no markings. Most likely the set was manufactured by another company, so it is possible that the same pieces may be found with other trade names.

Marked "A BEX moulding/Made in England," these pieces show the fluidity of line possible with Bakelite circa 1937 ($18-20 each). *Furniture and photograph from the collection of George Mundorf.*

A remarkable amount of detail was cast into these 1/2" to the foot metal pieces marked "Charbens/Made in England." Circa 1930s ($20-25 each). *Furniture and photograph from the collection of Patty Cooper.*

The opening door of this cast metal stove is marked "DCMT/Crescent." It is a large 3/4" to the foot in scale. This example appears to have been repainted ($25). *Stove and photograph from the collection of Patty Cooper.*

English Dollhouses and Furniture 73

Circa late 1880s tinplate furniture by Evans & Cartwright is a rare find for the collector. The clock and table both have a typical pressed design and faux mahogany finish. The table is approximately 1.25" to the foot in scale and the clock is much larger ($200+). *Furniture from the collection of Dian Zillner. Photograph by Suzanne Silverthorn.*

Metal bathroom pieces, sold under the tradename "Fairylite," are shown in both cream and green. All of the pieces are in a small 3/4" to the foot scale. The cream colored set is missing the high tank for the toilet ($15-20 each). *Furniture and photo from the collection of Patty Cooper.*

Jacqueline

"Jacqueline" has been identified by Margaret Towner as the trade name of B.&S. Mfg. & Distributing Co. Ltd. They are believed to have sold sets of painted steel dining room and bedroom furniture. Although the furniture is very Art Deco in style, it was probably manufactured after World War II.

Kleeware

O. & M. Kleemann, Ltd. produced plastic dollhouse furniture during the late 1940s and into the 1950s. Their furniture was marketed under the trade name Kleeware in 1947. The early Kleeware furniture, which was 3/4" to the foot in scale, was manufactured with the same molds used by the Renwal company in the United States. At first, the furniture was made of hard plastic but a later line, in updated styles, was produced in a softer plastic. The company also marketed a line of smaller furniture, 1/2" to the foot in scale.

The "Jacqueline" living room and dining room furniture is made of painted steel, mostly cut from flat pieces and folded into Art Deco designs ($15-20 each). *Furniture and photograph from the collection of Patty Cooper.*

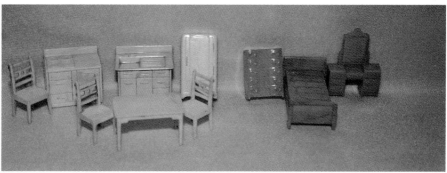

Made of hard plastic, these Kleeware kitchen and bedroom pieces are in a small 1/2" to the foot scale ($8-12 each). *Furniture and photograph from the collection of Ray and Gail Carey.*

74 English Dollhouses and Furniture

The 3/4" to the foot Kleeware pieces are indistinguishable from the American Renwal pieces made from the same molds, except that these are clearly marked "Kleeware/Made in England." There are no operable parts ($10-14 each). *Furniture and photograph from the collection of Patty Cooper.*

Meccano Ltd.

Best known as the maker of Dinky Toys, which included a large line of cast metal cars and trucks, Meccano was located in Liverpool and conducted business from 1901 until 1981. During the 1930s they produced very small scale, approximately 1/2" to the foot, cast metal furniture which was sold to furnish a cardboard "Dolly Varden" dollhouse. Four rooms of furniture were available including a bathroom, bedroom, dining room, and kitchen, but no living room. The furniture was enameled in bright colors and despite its small size, many of the pieces had opening doors or drawers. Unfortunately, the furniture was made of a metal alloy which deteriorated over time and pieces found in the United States are usually not in good condition.

Pit-a-Pat

E. Lehman & Company of London produced a line of wood dollhouse furniture under the tradename Pit-a-Pat from 1932 until World War II. The furniture varied in scale from 3/4" to 1" equals a foot. The company advertised that "well over a hundred articles, all perfectly produced" were included in the range. Although it is said to be fairly easy to find in England, the furniture was apparently not exported to the United States at the time it was manufactured. The Pit-a-Pat styles were typical of the 1930s middle class, rather chunky and comfortable looking, with velvet or faux leather upholstery on the sofas and chairs. The living room furniture included a sofa bed, known in England as a "put-u-up." Many of the pieces, including those for the dining room, were Tudor in style. Some had a rather distinctive beaded molding. The kitchen furniture was either dark stained or painted green or white. The nursery pieces were painted white and decorated with decals of children playing. The wide range of accessories included a tea tray, mantle clock, dinner gong, table radio, cutlery set, toilet paper, and a rather extensive selection of books. Pit-a-Pat used both ink stamps and paper labels to identify their products, and fortunately for today's collectors, quite a few of the labels have survived.

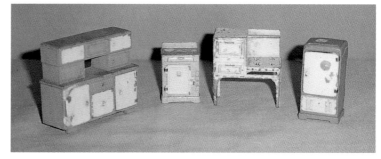

Dinky Toys furniture was produced by Meccano Ltd. of Liverpool to furnish a cardboard Dolly Varden dollhouse circa 1930s. The cast metal furniture was made in approximately 1/2" to the foot scale. The kitchen cabinet, electric cooker, and refrigerator all had opening drawers ($25-30 each). They are shown with a Tootsietoy stove to provide an indication of scale. *Furniture and photograph from the collection of Sharon Unger.*

The Dinky Toys bathroom pieces, by Meccano Ltd., are slightly smaller than standard 1/2" to the foot scale. The sink, with its deteriorated base, is typical of the poor condition in which these pieces are usually found ($25-30 each in good condition, significantly less as shown). *Furniture and photograph from the collection of Patty Cooper.*

The Pit-a-Pat living room chairs are covered in an imitation leather paper ($20 each). The wooden bed, in 3/4" to the foot scale, has wire mesh springs ($18-20). The stove has an opening oven door and appears to have been repainted ($15). *Furniture and photograph from the collection of Patty Cooper.*

English Dollhouses and Furniture 75

The bottom on the Pit-a-Pat bed has the paper label intact. *Furniture and photograph from the collection of Patty Cooper.*

Three Pit-a-Pat nursery pieces, circa 1930s, show decals of children playing. The wardrobe is 3.5" tall ($15-20 each). *Furniture and photograph from the collection of Ray and Gail Carey.*

A small scale (approximately 1/2" to the foot) cooking range of pressed tin is unmarked, but distinctly British in style, circa late 19th century ($75). *Stove and photograph from the collection of Patty Cooper.*

Made of cast iron, this English cooking range was meant to have a wood frame and may originally have been built into a dollhouse ($100-125). *Stove and photograph from the collection of Patty Cooper.*

Another pressed tin kitchen range is prominently marked "Made in England." This one seems newer and is approximately 3/4" to the foot in scale ($45). *Stove and photograph from the collection of Patty Cooper.*

GERMAN DOLLHOUSES AND FURNITURE

MORITZ GOTTSCHALK

The firm of Moritz Gottschalk was located in Marienberg, Saxony, and produced a wide variety of toys over a period of approximately seven decades. The Gottschalk company dominated the dollhouse market from the 1880s until the late 1930s when, with the advent of World War II, toy production in Germany was changed forever. Unlike Rufus Bliss, who died long before the first dollhouse was ever produced by the Bliss company, Moritz Gottschalk is believed to have been a talented artist who designed the early Gottschalk dollhouses himself and worked directly with the company's other artists until his death in 1905.

When these houses were first discussed in the early dollhouse books and in the pages of *International Dolls' House News*, it was believed that they might have been made in France. Collectors referred to the lithographed paper over wood dollhouses as Blue Roofs because, with a few exceptions, the roofs were painted blue. This is still a common designation and may be somewhat confusing to a new collector who encounters a Gottschalk house which has a brown lithographed paper roof, yet is being called a Blue Roof. Other dollhouses, which appeared to be from a different series, with painted exterior walls, window mullions of die-cut cardboard, and red painted roofs, were dubbed Red Roofs.

In 1980, Evelyn Ackerman published her book *Victorian Architectural Splendor in a Nineteenth Century Toy Catalogue*. This book reproduced pages from a German manufacturer's agent which clearly showed the houses known as Blue Roofs. Although the catalog did not give the name of the manufacturer or a date, it proved the German provenance of the houses. Interestingly, this catalog was found among papers in the estate of Albert Schoenhut, implying at least a tangential connection between the two companies, which will be discussed later. Over the years that followed, various researchers observed the common features of the Blue Roof and Red Roof houses. In 1982, British researcher Margaret Towner discovered that both series were produced by the firm of Moritz Gottschalk.

The Gottschalk company could be considered the world's most prolific manufacturer of dollhouses. The dollhouses were produced for such a long period of time, in great quantities, and with so many variations that the number of different models appears to be limitless. It is safe to say that Gottschalk produced a greater variety of houses than Bliss and Schoenhut combined. Evelyn Ackerman's book *The Genius of Moritz Gottschalk*, probably the most thorough published study of a single dollhouse manufacturer, includes over 120 different examples, yet even so, other designs continue to be found. The company also produced other structures of interest to dollhouse collectors including shops, rooms, fortresses, warehouses, theaters, stables, and later, garages. Some of the houses were specifically produced for export, including many in a style designed to appeal to the French market or printed with English words such as "With Lift." They were sold in upscale department stores and are often found with labels from F.A.O. Schwarz or other retailers. Some of the smaller, later examples have been found in more proletarian sources such as the Montgomery Wards Catalog.

This four-room Gottschalk Blue Roof features a symmetrical facade with three dimensional, rather than lithographed, door and window trim. The central peaked tower and windows with rounded tops show this to be one of the earliest Gottschalk models. The elaborate molding around the roof edge, bay windows, turned posts, and bay windows make this an especially appealing house (House too rare to establish price). 39.5" high x 30.25" wide x 20.25" deep. *House from the collection of Paige Thornton. Photograph by Steve Thornton.*

German Dollhouses and Furniture

The first dollhouses produced by the Gottschalk company were lithographed paper over wood, probably pine. Known to collectors as Blue Roofs, these houses were manufactured from the mid-1870s until the 1920s. In addition to the extensive detail achieved through lithography, the houses, especially early ones, often had pierced metal railings, wood moldings, and turned pillars, finials, and balustrades. The architectural structure of the houses ranged from rather simple boxes to elaborate structures with gables, towers, and offset porches. They ranged in size from less than a foot to over 40" tall. The interiors were papered in small prints, often with borders around the edges of the walls. Floor papers were printed in intricate parquet patterns. Larger houses featured interior doors with pewter handles and some houses were equipped with elevators. The front doors were rather distinctive, usually green with lines of paint used to give the illusion of panels. When found today, Blue Roof houses often have a crack in the back which may have occurred long ago. When the Bliss company compared their dollhouses to imported models that were "unsatisfactory in every way" because they soon warped and cracked, they were probably referring to Gottschalk, their main competition and probable inspiration.

Before the first World War, the Gottschalk company began to redesign its dollhouses, changing from the lithographed paper exteriors to more simplified painted ones which reflected modern taste, and perhaps were more economical to manufacture. Die cut pressed cardboard replaced the glass windows as well as the pierced metal railings. The predominate color of the roofs changed from blue to red, the source of the nickname "Red Roofs" used by today's collectors. Exteriors were usually painted a creamy yellow, although sometimes they were mustard yellow or had a green or brownish tint. The front doors continued to be the green ones with painted lines suggesting panels. The interior wall and floor papers were also similar to the ones used earlier, although a thorough study could probably document a more detailed evolution. The Red Roof dollhouses were offered in great variety, ranging from one room cottages to multi-storied houses with as many as ten rooms and hallways. They were often equipped with the latest conveniences including elevators, garages, and toilets which were usually located in an entryway closet.

Gottschalk houses were usually marked on the bottom with a four digit number. Researchers have been very interested in studying the numbers in order to designate specific models and try to determine dates. Some Blue Roof dollhouses have been found with numbers as high as the 4500s and Red Roof dollhouses with numbers starting around 4600. Usually penciled by hand, the numbers are often difficult to decipher and this may have created some false clues. For the collector who would like to try to determine where a particular house may fall in the Gottschalk sequence, Evelyn Ackerman's books are a helpful resource. The catalog reproductions show the same numbers in both script and print. A thorough study may help the collector to begin to decipher the handwritten numbers. Usually a "1" has a serif, a "7" has the continental cross, a "2" and a "4" may look very similar, and a "0" may have a loop at the top which makes it look like a "9". However, collectors should be aware that the same models were produced for long periods so numbers found on the bottom of dollhouses can not be used precisely to determine dates of manufacture. *The Universal Toy Catalog of 1924 (Der Universal Spielwaren Katalog)*, reprinted by New Cavendish Books, shows four different houses with the numbers 5602, 5619, 5992, and 6051, a wide variation in a single year.

One of the most elaborate of the Gottschalk Blue Roofs, this circa 1890 dollhouse features highly detailed turned and applied ornamentation. It contains four large rooms and two halls. It opens at both sides and the front. This house would have been produced only in very limited quantity and expensive even when new (House too rare to establish price). 53" high x 46" wide x 32" deep. *House from the collection of Anne B. Timpson. Photograph by Mary Kaliski courtesy of Miniature Collector Magazine.*

This very rare, early Blue Roof, with an unknown number, has two separate entrances. It is symmetrical in style with a central gable and projecting, second story bay windows on each side. It is all original inside and out, with silk curtains at each window. 32" high x 34" wide x 21" deep (House too rare to determine price). *House from the collection of Anne B. Timpson. Photograph by Nick Forder.*

A more modest early example, Gottschalk No. 2793 is symmetrical and features turned posts and applied window trim over a pattern of lithographed brick. The front gable contains an unusual cut-out ornament. There are no windows on the sides ($1800-2000). 22.5" tall (to top of roof) x 15" wide x 12" deep. *House from the collection of Madeline Large. Photograph by Patty Cooper.*

many" on the bottom along with the penciled number 5175. There is nothing printed or written on the bottom of the Schoenhut. Both houses have the green door, with lines indicating panels, so typical of Gottschalk. The Gottschalk influence can easily be seen in other, later Schoenhut houses, with their cream colored exteriors, red roofs, and pressed cardboard mullions.

The circa 1885 Gottschalk catalog reproduced by Evelyn Ackerman in her book *The Genius of Moritz Gottschalk* states that furniture was available for their dollhouses, although none is shown except in the kitchens and shops. There is no way to know if this furniture was actually produced by Gottschalk or supplied by another manufacturer. However, by 1924, *The Universal Toy Catalog* advertised several boxed sets of furniture identified as being made by manufacturer number 7, which was clearly Gottschalk. Most of the furniture shown in the catalog was made of die cut pressed cardboard with a glossy enamel finish, often trimmed in gold, the same technique used in making the windows of Red Roof dollhouses. The furniture was produced in several different scales, ranging from approximately 1/2" to the foot to a large 1" to the foot. Thus, it was possible to furnish any size Gottschalk dollhouse with the appropriate furniture. The larger Gottschalk pieces were often made of wood rather than pressed cardboard, but were in the same style and had the same finish. The *Universal Toy Catalog* advertised sets of drawing room furniture in three different sizes, all in a mahogany color. Three sizes of white enameled bedroom sets were offered. However, the catalog contained only one kitchen set and one bathroom set, both in white, and presumably in the medium size.

There are indications that the Schoenhut Company may have had a connection to the firm of Moritz Gottschalk. Two houses have been found, one marked Schoenhut and the other obviously a Gottschalk "Red Roof", which are so similar that it must be assumed they were manufactured by the same company. The Schoenhut house has the same paper label as the early faux stone Schoenhuts. The label states "Made in USA/ The A. Schoenhut Company/Philadelphia, PA." This house is so clearly identical in construction to the Gottschalk that it belies the label. It is difficult to think of a reason for Schoenhut to have made such an exact copy. It seems much more likely that the shell was imported from Germany and finished in Philadelphia. The floor papers are very typical of Schoenhut with printed wood planks as opposed to the Gottschalk parquet. The wallpapers contain small geometric patterns not seen in any known Gottschalk houses, but similar to paper found in other Schoenhuts. The side windows on the Schoenhut house are actually printed decals which have also been found on later Schoenhut colonial style houses. The nearly identical German house has typical Gottschalk floor and wallpapers and stenciled windows on each side. The houses are exactly the same size, but there are a few other minor differences. The stairway in the Schoenhut house does not have an opening at the top, whereas in the Gottschalk version, a doll can actually go up the stairway into the attic. There is no toilet in the closet of the Schoenhut version and there is nothing, no nail holes or tears in the floor paper, to indicate it ever contained one. The Gottschalk has the printed word "Ger-

The smallest of the Blue Roof dollhouses shown, Gottschalk No. 3035 features an unusual exterior of lithographed clapboard siding. Three turned posts support the roof of the porch, but the other architectural features, such as the balustrade, window trim, and gable decoration are printed rather than applied ($1200-1500). 13.5" high x 9" wide x 8.25" deep. *House and photograph from the collection of Patty Cooper.*

Another line of pressed cardboard furniture was produced in Germany in a design meant to imitate wicker garden or porch furniture. Although these pieces employed the same material and method used in the Gottschalk furniture, the garden pieces have not been found in any catalog, so can not be definitely attributed to that company. The faux wicker pieces were produced in scales ranging from approximately 1/2" to the foot to nearly 1 1/2" to the foot. They were made in a variety of colors including orange, green, and cream, often with contrasting borders of red, green, or blue. Of the pressed cardboard furniture, the garden sets are somewhat easier for today's collector to find, suggesting that they were made later or for a longer period of time.

Gottschalk No. 3548, circa 1890s, is one of a series, numbered in the 3000s, which features printed half-timbering and alternating bands of red and brown brick or boards. This is one of the most elaborate, with a bay window and total of four porches. The third story dormer is especially interesting because it contains a door which provides access to the attic. Pierced metal railings decorate all of the porches. The original silk curtains, in typical colors of blue and pink, can be seen at the windows. The house opens in two sections and contains four rooms (House too rare to determine price). 25" high (not including chimneys) x 20.5" wide x 16" deep. *House from the collection of Anne B. Timpson. Photograph by Mary Kaliski courtesy of Miniature Collector Magazine.*

Gottschalk No. 3582 has had its roof repainted, but is still clearly a Blue Roof. Even without knowing its number, the more exaggerated posts and decorative chimney caps would suggest that it is slightly earlier than the following example. The pierced metal railing with its circular design is commonly found on dollhouses in this series ($2500-3000). 21" high (to roof) x 17.5" wide x 11" deep. *House and photograph from the collection of Patty Cooper.*

80 German Dollhouses and Furniture

The simple two-room interior of No. 3582 is something of a disappointment when compared to the elaborate facade. The house is furnished with Gottschalk pressed cardboard furniture. *House and photograph from the collection of Patty Cooper.*

Gottschalk No. 3583 opens in two sections to reveal four rooms with original wallpapers. There are two functional interior doors. Similar houses, with model numbers varying by the final digit, have been found in slightly different sizes and opening from the sides. *House and photograph from the collection of Patty Cooper.*

Gottschalk No. 3583 has slightly narrowed turned posts and a pierced metal railing in front of the porch and upstairs balcony ($2800-3200). 22" high x 20" wide x 10" deep. *House and photograph from the collection of Patty Cooper.*

German Dollhouses and Furniture 81

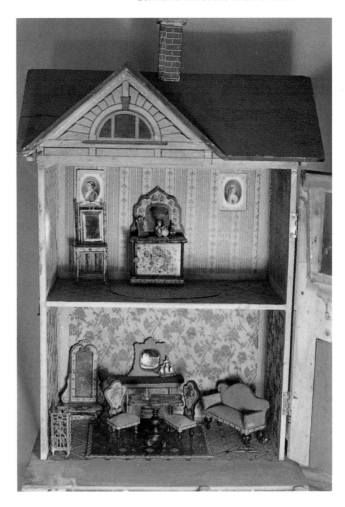

The two-room interior of No. 3868 is best suited for small furniture in 1/2" to foot scale. Note that the bedroom wallpaper is the same as that shown in No. 3582. The furniture is lithographed German with some soft metal pieces. *House and photograph from the collection of Patty Cooper.*

Although part of this same series, Gottschalk No. 3868 is a modest house, more affordable both at the turn of the century and today. This house has lithographed windows on the side and in the off-center front gable. Architectural interest is created by the second story balcony, with its simple pierced metal railing, and printed stained glass window flanked by two cut-out windows ($1800-2000). 19" high x 12" wide x 7.75" deep. *House and photograph from the collection of Patty Cooper.*

With its asymmetrical front gable and lithographed half-timbering, this house resembles those in the 3000 series. But it is also a link to the transitional houses which follow. The exterior of the house has a flat finish, resembling stucco, instead of the printed bricks and there are lithographed shutters. The two columns which support the porch roof are no longer turned, but made of squared lumber, notched and highlighted with stripes of red and black paint. The pierced metal railing of the balcony is in a different design and the door, although it has the usual painted outlines, is red rather than green. The number of this house is unknown, but would probably be in the 4200s ($2500-3000). *House and photograph from the collection of Anne B. Timpson.*

A significant departure from earlier styles, this Gottschalk house could be considered the beginning of the change from Blue Roofs to Red Roofs. This house, and its smaller sister, are literally neither. Instead, their roofs are covered with lithographed paper in a pattern of diamond-shaped shingles. The exterior papers are a drastic change. Although there are still printed architectural details, including door and window trim, the wall surfaces are unpatterned to simulate stucco. The flowing roof lines of the front and side parapeted gables show the influence of the Art Nouveau style. These flowing lines are echoed in the painted mullions of each window. The house opens on each side and contains four rooms. There are also two very small, but accessible attic rooms. The interior stairway has an elegantly curved metal railing. The penciled number on the bottom is illegible, but it is believed that it should be in the 4000s. It contains a circular paper label from F.A.O. Schwarz (Not enough examples to determine price). 36" high x 27" wide x 17" deep. *House from the collection of Anne B. Timpson. Photograph by Nick Forder.*

German Dollhouses and Furniture 83

Gottschalk No. 4450 is much smaller, but clearly related to the previous example. Both feature lithographed roof papers, peaked towers, and parapeted gables. This example has printed shutters at each window ($1600-1800). 18.5" high x 14" wide x 9" deep. *House and photograph from the collection of Patty Cooper.*

This unmarked, two-room house is typical of a series that has been known to collectors as "Deauville." Originally believed to be French, there is now a general consensus that these houses were made in Germany. One theory suggests that they were made by Gottschalk for the French market. The decorative railing and tower roof have been replaced ($800-1200). 22" high x 13" wide x 8.5" deep. *House from the collection of Dian Zillner. Photograph by Suzanne Silverthorn.*

Access is provided to all three rooms of No. 4450 through openings at the right side, front, and in the front gable. The only access to the tower room is through the attic, a very tight squeeze for an adult hand. The house is appropriate for small furniture approximately 1/2" to the foot in scale. *House and photograph from the collection of Patty Cooper.*

84 German Dollhouses and Furniture

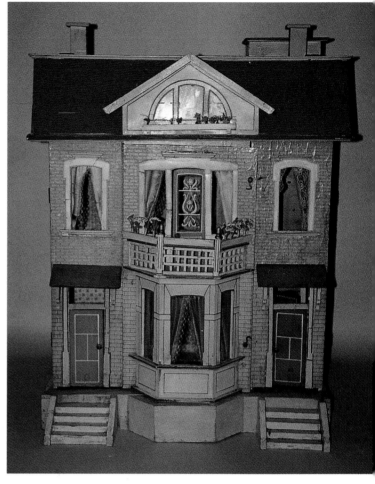

Gottschalk No. 4688 is also part of the transition from Blue Roofs to Red Roofs. It still has the lithographed paper which defines the Blue Roof series, but clearly has a red roof and a die-cut pressed cardboard, as opposed to pierced metal, railing. This house presents a symmetrical facade with two separate entrances ($2800-3000). 27.5" high x 23" wide x 14" deep. *House and photograph from the collection of Lois L. Freeman.*

Gottschalk no. 467? (the last number is illegible) contains four rooms and an attic accessible through a hinged door in the front gable. A very similar house was shown in a 1922 advertisement. Note the decorative stenciling in the peak of the gambrel-roofed gable and on the porch supports ($2000-2200). 26" high (not including chimneys) x 25" wide x 13" deep. *House and photograph from the collection of Lois L. Freeman.*

The inside of Gottschalk no. 467? features two sets of two-story bay windows which open with the hinged fronts of the house. The wallpapers are original and the tiled kitchen paper should be noted. *House and photograph from the collection of Lois L. Freeman.*

German Dollhouses and Furniture 85

Gottschalk no. 5182 is a small one-room cottage with an attic. There are two doors opening onto the wrap-around porch. A side door provides access to the attic ($750-850). 14" high x 13" wide x 8" deep. *House from the collection of Dian Zillner. Photograph by Suzanne Silverthorn.*

The front of No. 4688 opens in one section and the roof also lifts up to reveal three rooms and three hallways which provide access to the highly desirable elevator. The working elevator is operated by a crank which is attached to a pulley. The door of the wooden elevator opens so that dolls may actually ride. This house is large enough to easily accommodate furniture in 1" to the foot scale. *House and photograph from the collection of Lois L. Freeman.*

Gottschalk no. 5182 was advertised in a 1923 Montgomery Ward catalog. The house sold for $3.89 furnished with two straight chairs, a settee, a table, and a music cabinet, all made by Gottschalk of pressed cardboard. *Ad from the collection of Dian Zillner. Photograph by Suzanne Silverthorn.*

86 German Dollhouses and Furniture

Gottschalk No. 518(6) is a simple two room cottage with a gambrel roof. The attic contains one large room, accessible through doors on each gabled end ($1000). 16.5" high x 19" wide x 12" deep. *House and photograph from the collection of Patty Cooper.*

Gottschalk No. 5465 has porches which extend the full width of both the first and second stories. There are two sets of French doors opening on to the verandahs. Access to the attic is through a hinged door in the front gable and a trap door on the left side of the roof. Inside are two large rooms and two halls with a staircase. The first floor hallway also contains a closet which houses the toilet. Ceilings are approximately 9" high, so the house can accommodate furniture which is almost 1" to the foot in scale ($2000-2200). A very similar house with a side porch, no. 5466, was shown in a 1922 advertisement. 26" high (not including chimney) x 25" wide x 13" deep. *House and photograph from the collection of Patty Cooper.*

No. 518(6) is an example of the most simple of Gottschalk interiors, with no stairway or built-in toilet. It is interesting to compare it to No. 5175, a smaller house in a similar style which contains a more detailed interior. The pressed cardboard furniture is Gottschalk. *House and photograph from the collection of Patty Cooper.*

No. 5465 contains four rooms and an attic. One of the rooms is the front entry which contains a stairway and closet with toilet. The wallpaper in the kitchen has been replaced. Most of the furniture is Gottschalk and the dolls are German. *House and photograph from the collection of Patty Cooper.*

German Dollhouses and Furniture 87

There is no number on the bottom of this Gottschalk house, only the words "Germany" and "F.A.O. Schwarz." The gambreled front gable and shape of the die-cut pieces in the porch rail suggest that it is a close relative of no. 467? and no. 5465, therefore circa 1922. This house features a porch which extends across its front and wraps around the side. Inside, there are three rooms and an attic accessible through the hinged door in the front gable. It does not contain a stairway or toilet ($1500-1800). 23" high x 21" wide x 12" deep. *House and photograph from the collection of Ruth Petros.*

Gottschalk no. 5472 resembles a Swiss chalet. The smallest of the houses shown, it contains one room plus an attic which is accessible through a hinged door in the back. The ceiling is only 4.25" high, making the house appropriate for furniture 1/2" to the foot in scale. The house features a small walled yard with trellis ($500-600). 13.5" high x 13" wide (at base) x 11" deep (at base). *House and photograph from the collection of Patty Cooper.*

Gottschalk no. 5600 has a symmetrical facade with double gables, two bay windows, and a second story verandah which spans the width of the house. The front opens in one piece and the hinged roof can be lifted to provide complete access to all three floors. The base is incised to represent stone ($1800-2000). 19.5" high x 17.5" wide x 9.25" deep. *House and photograph from the collection of Patty Cooper.*

Gottschalk no. 5600 contains six rooms, at least two of which could be considered hallways. The room on the lower right contains both a staircase and a closet which holds the toilet. There are two interior doors. The ceilings are approximately 6" tall so this house is most appropriate for furniture 3/4" to the foot in scale. All of the wallpapers are original. *House and photograph from the collection of Patty Cooper.*

Gottschalk no. 5694 is another example with a "flip-top" which allows complete access to the space under its gambrel roof. The die-cut pressed cardboard mullions appear to be missing from two of the second story windows ($1400-1600). 20" high x 26" wide x 11" deep. *House and photograph from the collection of Lois L. Freeman.*

With its two front sections opened and roof hinged up, Gottschalk no. 5694 has many desirable features for its size: four rooms, a stairway, built-in toilet, interior doors, two porches, and original wallpapers. *House and photograph from the collection of Lois L. Freeman.*

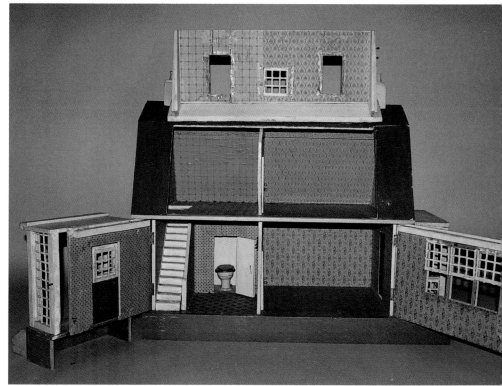

German Dollhouses and Furniture 89

Although no penciled numbers, only the words "Made in Germany," could be found on the bottom of this gambrel-roofed house, its similarities to no. 5694 are obvious. A highly desirable house with two porches, this house also features a garage. At first glance, this house might appear to be missing its base, but comparison with other Gottschalk houses with garages suggests that these houses were designed so that a car could be "driven" into the garage without having to climb over a curb ($1800-2000). 17" high x 22" wide x 11" deep. *House and photograph from the collection of Carol Miller.*

A larger Gottschalk house No. 6163 also features a garage. It has a two story bay window, two-story side porch, and flower boxes. The one piece front must be opened to access the attic through the hinged door in the front gable. Inside are five rooms, including attic and hallways. There is a stairway and built-in toilet ($2600-2800). 26" high x 32" wide x 16" deep. *House from the collection of Lisa Kerr. Photograph by Bruce Kerr.*

Gottschalk No. 5992 is illustrated in *The Universal Toy Catalog* of 1924, a reprint of a German toy catalog. The front of the house opens in one piece, taking most of the off-center porch with it. There is one main room and an attic space accessible through an opening in the back ($800-900). 15.5" high x 15.75" wide x 9.5" deep. *House and photograph from the collection of Marilyn Pittman.*

No. 6168 is a Tudor style mansion containing an impressive total of ten rooms, including those in the attic, basement, and the conservatory. It also has a two-car garage, not visible in the photograph, accessible through sets of double doors at the right rear of the house. The front facade opens in two asymmetrical sections. Inside, there is a stairway with Craftsman-style balustrade and a closet with a toilet. The kitchen features a working dumbwaiter. Servants' quarters are located in the basement on the same level as the garage. This large house easily accommodates 1" scale furniture. The only other known example is in the collection of the Kansas City Toy Museum (Too rare to establish price). 46.5" high x 49.5" wide x 25" deep. *House and photograph from the collection of Anne B. Timpson.*

Although outwardly this unmarked Red Roof is rather simple, when the front is opened and the dormer roof lifted, there are four playable rooms and a closet for the toilet. The house contains two interior doors as well as an archway which provides access to an alcove in the attic. *House and photograph from the collection of Marilyn Pittman.*

The bottom of this house is stamped "Made in Germany" in purple ink, with no number. A similar example, but without the side porch, was shown in the Spring 1994 issue of *International Dolls House News* and said to have the number 6214 penciled on the bottom, dating it circa mid-1920s. The "H"-shaped pressed cardboard pieces in the porch rail relate it to No. 5784 and its textured red roof is the same as No. 6309. All three of these houses have "flip-top" roofs, although only the dormer roof opens on this example ($1200-1400). 15" high x 18.25" wide x 10" deep. *House and photograph from the collection of Marilyn Pittman.*

German Dollhouses and Furniture 91

Gottschalk No. 6309 has two features not seen in earlier Red Roof houses. The balusters in the porch rail are turned rather than made of die-cut pressed cardboard and the red roof is textured to imitate shingles ($2100-2300). 24.5" high x 21" wide x 13.5" deep. *House and photograph from the collection of Patty Cooper.*

The front of No. 6309 swings open in one piece and the roof flips up. Each of the three floors contains one large room and a hallway which opens onto the elevator shaft. The elevator, a highly desirable feature in a Gottschalk house, is operated by a pulley connected to a hand crank located at the right rear of the house. Next to the elevator, on the first floor, is a closet which contains a toilet. Operable doors connect the rooms and provide access to the elevator. Other doors, with pewter, butterfly-shaped handles, open onto the side porches. *House and photograph from the collection of Patty Cooper.*

Gottschalk No. 6337 is Mediterranean in style with a hipped roof. French doors open onto the extensive verandahs. The house is appropriate for furniture in 3/4" to the foot scale ($2000-2400). 23" high x 25" wide x 15.5" deep. *House and photograph from the collection of Lois L. Freeman.*

92 German Dollhouses and Furniture

The front of Gottschalk No. 6337 opens in two sections and there is a hinged door in the front gable which provides access to the attic. Inside, there are four rooms in addition to the attic and two hallways. There is a staircase and a closet with toilet. All of the papers are original and appropriate to the function of the rooms. *House and photograph from the collection of Lois L. Freeman.*

Gottschalk No. 6373 has a deceptively plain, flat facade with windows which are created by decals instead of being cut out. The side porch features turned balusters typical of later Gottschalks ($2000-2400). 21.5" high x 20" wide x 10.75" deep. *House and photograph from the collection of Anne B. Timpson.*

The simple exterior of No. 6373 conceals a wonderful surprise. The front facade folds down to create an attached garden complete with pressed cardboard fence and flower bed. A gate in the fence opens to invite a doll to enter. With the roof flipped up, there is access to three large rooms and three hallways. A stairway is provided as well as a closet with toilet. *House and photograph from the collection of Anne B. Timpson.*

Gottschalk No. 666? (last number illegible) has a dignified, symmetrical facade. It features a hipped roof, twin windowless gables, and a second story balcony. Inside, the house has two rooms plus the attic, a stairway, and closet for the toilet ($1600-1800). 22" high x 20.25" wide x 13.75" deep. *House and photograph from the collection of Ray and Gail Carey.*

The wall and floor papers inside Gottschalk No. 5175 are typical of that company. The house opens in two sections to reveal two rooms and a central hallway. The stairway leads to an opening at the top through which a doll could enter the attic. The closet behind the stairs contains a toilet. *House and photograph from the collection of Patty Cooper.*

Gottschalk No. 5175 is pictured out of sequence in order to facilitate comparison to the nearly identical Schoenhut. Access to the attic is through hinged doors on each side of the house. The facade is decorated with pressed cardboard lattice and a window box containing its original paper flowers. The windows are glass and appear to be replacements ($1000-1200). 15" high x 16" wide x 10.5" deep. *House and photograph from the collection of Patty Cooper.*

This marked Schoenhut house is so clearly identical to Gottschalk No. 5175 that it seems obvious that it, too, originated in Germany ($1000-1200). 15" high x 16" wide x 10.5" deep. *House and photograph from the collection of Patty Cooper.*

The side of the Schoenhut house contains a paper label which states "Made in USA/The A. Schoenhut Company/Philadelphia, Pa." This is the same type label used on the faux stone Schoenhuts circa 1923. *House and photograph from the collection of Patty Cooper.*

The inside of the Schoenhut house reveals differences which suggest that although the house was probably built in Germany, it was finished in America. The floor papers and wallpapers are typical of the Schoenhut company. The stairway ends at the ceiling and the closet is only a closet, with no evidence that it ever contained a toilet. *House and photograph from the collection of Patty Cooper.*

The gray stucco house contains one room and an attic. It is furnished with pressed cardboard faux wicker furniture. One other variation of this house, with a garden on the right, has been seen. It, too, had no number or other identification. *House and photograph from the collection of Nancy Roeder.*

This unusual house is included here for lack of a more appropriate categorization. Although it bears some Gottschalk characteristics, such as the pressed cardboard windows and outlined garage doors, it also has many features atypical of that company, including its gray stucco exterior. Some notable attributes are the garage, patio with arched doorway, and flip-top dormer. The bottom of the house is stamped "Made in Germany" in purple ink ($1800-2000). 16.5" high x 22.5" wide x 10" deep. *House and photograph from the collection of Nancy Roeder.*

German Dollhouses and Furniture 95

A Gottschalk kitchen is shown with furniture presumably made by that company. Many of the accessories have been replaced ($1800-2500). 12.75" tall x 28.5" wide x 12.5" deep. *Kitchen from the collection of Madeline Large. Photograph by Patty Cooper.*

The smallest Gottschalk pressed cardboard furniture is approximately 1/2" to the foot in scale. In addition to the pieces shown, a bed and a table are known to have been made in this size. The small furniture is the most difficult for today's collector to find ($15-30 each). *Furniture and photograph from the collection of Patty Cooper.*

Some of these 3/4" to the foot scale pieces are shown in *The Universal Toy Catalog of 1924*. The shiny reddish brown enamel is described in the catalog as mahogany. The original set would have included two more chairs, a pier mirror, and a framed picture. The bentwood type rocker is not shown in the catalog (Straight chair $25-30, other pieces $35-40 each). *Furniture and photograph from the collection of Patty Cooper.*

These pieces are only fractionally larger, a little over 3/4" to the foot in scale. This is the most commonly found size and is appropriate for many of the Gottschalk houses. The cabinet doors are operable and the "upholstery" is flocked paper (Cabinet $50, other pieces $35-40 each). *Furniture and photograph from the collection of Patty Cooper.*

96 German Dollhouses and Furniture

Another pressed cardboard set in a large 3/4" to the foot scale has padded seats upholstered in silk (Chairs $25-30, other pieces $35-40 each). *Furniture and photograph from the collection of Patty Cooper.*

This Gottschalk set, in large 3/4" to the foot scale, includes a different cabinet and a settee with a cut-out back (Chairs $25-30, other pieces $35-40 each). *Furniture and photograph from the collection of Patty Cooper.*

This set in 1" scale shows a cabinet with double doors. The chairs were available both with and without padded seats (Chairs $25-30, desk $45, table $45, sofa $50, cabinet $50). *Furniture and photograph from the Patty Cooper.*

This slightly larger Gottschalk set is also 1" to the foot in scale. The seats are padded and some of the parts are wood rather than pressed cardboard (Chairs $25-30, desk $45, table $45, sofa $65, cabinet $60). *Furniture and photograph from the collection of Ray and Gail Carey.*

German Dollhouses and Furniture 97

The triangular back of this 1" scale sofa ($125) allowed it to fit into an unusual octagonal house by Gottschalk. Yet another cabinet variation ($85-100) is shown along with a round table ($65-85). *Furniture and photograph from the collection of Roy Sheckells.*

The Gottschalk pressed cardboard cabinets were made in a variety of styles and sizes. All of the doors are operable. A Montgomery Ward advertisement from 1923 described these as "music cabinets." ($35-50 each). *Furniture and photograph from the collection of Patty Cooper.*

This white enameled bedroom set is 3/4" to the foot scale. The beds are 4.5" long. An original bedroom set would have included an armoire and framed picture, with only one bed (Chair $25-30, other pieces $45-50 each). *Furniture and photograph from the collection of Patty Cooper.*

Although not shown in any catalogs, the style and material of this fireplace suggest that it was made by Gottschalk ($45-50). *Furniture and photograph from the collection of Ray and Gail Carey.*

These pieces are slightly larger, with beds 4.75" long. A bedroom set would have contained only one bed, however, two are shown to illustrate that the design was sometimes cut-out rather than just impressed. The door of the armoire is operable. (Chair $25-30, other pieces $45-50 each). *Furniture and photograph from the collection of Patty Cooper.*

98 German Dollhouses and Furniture

Another group, believed to be Gottschalk, contains a wider bed which is 5.25" long. The cabinet may be a kitchen piece (Chair and nightstand $25-30 each, other pieces $45-50 each). *Furniture and photograph from the collection of Patty Cooper.*

Bedroom pieces were also made in 1" to the foot scale ($25-50 each). *Furniture from the collection of Madeline Large. Photograph by Patty Cooper.*

This grouping of 1" to the foot scale, white-enameled furniture includes pieces which may have been intended for the kitchen, bedroom, or garden (Chairs or nightstand $25-30 each, tall chest or cupboard $45-65, bench $60, table $35, shelf $45). *Furniture and photograph from the collection of Roy Sheckells.*

German Dollhouses and Furniture 99

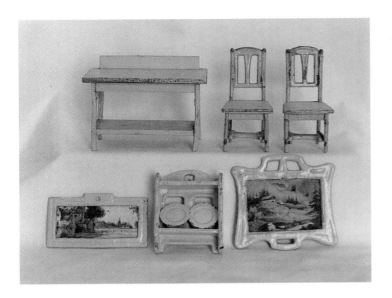

Kitchen chairs, a table, and hanging shelf in 3/4" scale are shown with two pictures in pressed cardboard frames ($25-30 each). *Furniture and photograph from the collection of Patty Cooper.*

This scroll-cut furniture in 3/4" to the foot scale was often sold in the "Deauville" type houses believed to have been produced by Gottschalk for the French market. It is possible that the houses were finished and furnished in France ($30-40 each). *Furniture and photograph from the collection of Ray and Gail Carey.*

Garden furniture with a wicker pattern was also made in pressed cardboard, using many different color combinations. It has been attributed to Gottschalk, but there is no proof of this. The furniture was made in at least three different sizes, including the 1/2" to the foot scale shown in the top row to the large 3/4" to the foot scale on the bottom row. The 1 1/2" to the foot scale is not shown ($40-50 each). *Furniture and photograph from the collection of Patty Cooper.*

This 1" scale dining room furniture, in orange enamel, bears a similarity to verified Gottschalk pieces. The backs of the chairs appear to be made of pressed cardboard and the table has the same construction as those shown earlier. The simpler, more modern lines of the chairs suggest that they may have been made at a later time than the mahogany pieces (Chairs $20, table $30, cabinet $35). *Furniture and photograph from the collection of Patty Cooper.*

CHRISTIAN HACKER

Christian Hacker of Nuremberg produced exquisite dollhouses, as well as kitchens, warehouses, and shops, from the mid-1800s through 1914. The company is best known for its three-story houses with mansard roofs. Because the houses were somewhat French in appearance, for many years they were believed to have been made in France and were described that way in many of the older dollhouse books. In the 1970s, one of these houses was discovered with the inscription "Christian Hacker, Nuremberg, Germany, 1875, Maker of wooden playthings." It was also stamped with a crown-topped shield containing the initials "CH" below the word "shutzmarke," which is German for trademark. This clue provided the link which allowed researchers to properly attribute many other styles to the Hacker firm.

Other than the trademark stamp, which seems equally likely to be missing or present, there is no one characteristic which applies to all Hacker dollhouses. Instead, there are common threads which have been used to connect unmarked houses to other houses or stores which have been verified as Hackers. As is true of any antique or collectible, familiarity with many different examples helps develop the ability to recognize similarities in finish, style, weight, or quality which may be difficult to photograph or even explain. Researcher Anne B. Timpson, who has had extensive experience with Hackers both as a collector and a dealer, has provided the following observations on elements which may be found in Hackers:

The bases usually had a distinctive beveled molding around the edge rather than just the plain, squared box base found on many other houses from the same period. Balustrades were either made of wood with a distinctive turning pattern consisting of two "vases" separated by a "ring," or of metal with a decorative pattern in the cut-out design. Windows in Christian Hacker houses usually had glass panes with mullions which were painted, often in blue. Inside, the windows were held in place by separate strips of wallpaper, which framed the glass on all four sides. Original curtains or draperies were sometimes held back by strips of ribbon secured with gilt embossed medallions. Curtains in later houses were gathered onto turned rods with finials at each end. Many of the doors had gilt striping and some were supplied with tin doorbells. When elevators were provided, they were operated by an elaborate clockwork mechanism rather than the simple crank used on Gottschalk houses.

Christian Hacker houses were designed for and exported to both England and the United States at the time they were originally manufactured. Many have been found with labels from F.A.O. Schwarz. The Hacker houses, like their Gottschalk counterparts, often had model numbers penciled underneath. But so far, no one has published a thorough study which establishes a sequence or links a date to these numbers. Generally, the Hacker houses were not mass produced to the extent that Gottschalk houses were. They were usually of a higher quality, with heavier doors and more expensive looking hardware, including cast metal, double flanged doorknobs, and beaded, electrified chandeliers. These houses were expensive even during the time they were being manufactured.

The mansard roofed houses, which are most easily identified as Hackers, were designed to be stackable, with each level, including the roof, resting on the one below. The front of each section usually opened separately. The roofs were papered in a distinctive diagonal slate pattern. The fronts of the houses and parts of the interiors were frequently embellished with elaborate transfer designs. The transfers were especially distinctive when used as ceiling medallions. The interior walls were papered in rich florals and stripes while the floors had intricate parquet designs. Two wonderful features found in many Hacker houses are built-in mantels and Welsh cupboards, painted cream, and trimmed with blue lines. Hacker kitchens, both in dollhouses and as separate rooms, usually had marbleized papers on the walls and diagonally printed tile in either red, blue, or black on the floor.

The Christian Hacker company made houses in many other styles including a series with hipped roofs and one with a formal garden. Hacker houses dating from the turn of the century have been found in the Queen Anne style, with turrets, porches, and rounded windows typical of high Victorian architecture. Most, although not all, of the Hacker houses were quite large in scale and very impressive. They ranged in size from two to four stories tall, with as few as three or as many as eight rooms. Although many Hacker houses have been identified, it is likely that many other models are still unattributed. Their size, rarity, and rich detail usually allows them to command very high prices with today's collectors.

This large, two-room kitchen is marked on the bottom with the Christian Hacker "shutzmarke," a crown-topped shield containing the initials "CH." Many of the furnishings are original ($800-1500). 14" high x 39" wide x 14.75" deep. *Room and photograph from the collection of Marilyn Pittman.*

This very rare, circa 1860 Hacker features a fenced garden with a tin fountain, swing, and gazebo. A cistern hidden in the attic originally supplied water to the fountain. The balusters in the fence have the distinctive "vase/ring/vase" turnings found on other Hackers. The house contains two long rooms and opens from the back (Too rare to determine price). 20" high x 22" wide x 21" deep. *House and photograph from the collection of Anne B. Timpson.*

This Queen Anne style Hacker is believed to be circa 1900. The house features a wrap-around porch, painted board and batten siding, and an elaborate tower. There are four rooms and two halls inside, with an elevator operated by an elaborate clockwork mechanism (House too rare to determine price). 53" high x 48" wide x 37" deep. *House and photograph from the collection of Anne B. Timpson.*

This circa 1880s house with a mansard roof is typical of the style most commonly associated with Christian Hacker. It contains six rooms with two halls and is constructed in three stackable sections. The kitchen still has the original built-in Welsh dresser and stove surrounds. The window mullions are painted on the glass windows and the double doors have faux-grained painted panels. The house was originally played with by children of a professor at the Groton School in Massachusetts (Too rare to determine price). 40" high x 35.5" wide x 18" deep. *House and photograph from the collection of Anne B. Timpson.*

This circa 1890-1900 Hacker house features two-story porches on the left and front. The distinctive "vase/ring/vase" pattern can be seen in the turnings of the second story balustrade. It also has a tin widow's walk, typical of some Hacker houses (House too rare to determine price). 34" high x 31" wide x 25.5" deep. *House and photograph from the collection of Anne B. Timpson.*

Another Queen Anne style house by Christian Hacker is covered with lithographed paper simulating brick and stone. The roof is painted blue with black trim (House too rare to determine price). 40" high x 27" wide x 25.5" deep. *House from the collection of Anne B. Timpson. Photograph by Nick Forder.*

The living room in the lithographed, Queen Anne style Hacker has the original wallpaper with ornate borders. *House from the collection of Anne B. Timpson. Photograph by Nick Forder.*

D.H. WAGNER & SOHN
AND GERMAN DOLLHOUSES OF UNKNOWN ORIGIN

There is little information available about Wagner dollhouses, which are infrequently found in the United States. It is known that the firm of D.H. Wagner & Sohn of Grunhainichen in Saxony was established in 1742. They also had facilities in Nuremberg and Sonneberg. They made dolls, metal and wooden toys, forts, and dollhouses. By the 1930s, they were being represented in London by agent Fred H. Allen Ltd. They produced dollhouses specifically for the English market and this has sometimes created confusion about their country of origin.

Several Wagner houses have been shown in the pages of *International Dolls House News* along with reprinted advertisements. These examples provide some clues to identifying a Wagner dollhouse, but none of the following characteristics apply to all known examples. The dollhouses seem to be mostly small in scale, between 1/2" and 3/4" to the foot. The windows may be made of pressed cardboard or metal and have square mullions. Some examples have lithographed windows with shutters, lace curtains, and t-shaped mullions which form a pattern of one small horizontal pane over two larger vertical panes. Several of the doors have a pattern of three vertical lines in the lower section. One of the most distinctive characteristics is a rather mottled, stenciled pattern of tile, brick, or stone, often found on the roofs, fronts, or bases. Stenciled half-timbering and stenciled shrubbery are other features commonly found on Wagner houses.

Found in parts by its current owner, this large Hacker features an unusual glassed conservatory in place of the second story porch found on the previous house. It has four main rooms, a central hall with stairway, and an attic. There are two small rooms attached to the back of the second story which create a bathroom and an alcove. A similar house was advertised in a 1913 Gamages catalog (House too rare to determine price). 43" high x 44.5" wide x 30" deep. *House and photograph from the collection of Ray and Gail Carey.*

Even the smaller Hacker houses are striking. This model contains three large rooms. The front opens in one piece and the front of the roof is removable. Underneath, the house is stamped "Made in Germany" with the handwritten numbers 504/2 and the initials "CH." The edges of the base are beveled as on the larger Hacker houses ($1800-2500). 21.75" high x 18" wide x 13.5" deep. *House from the collection of Madeline Large. Photograph by Patty Cooper.*

This Wagner house was found completely overpainted, but the owner is in the painstaking process of scraping it down to its original finish. The house contains two long rooms with the original wallpapers ($400-500). 16.5" high x 16.5" wide 10" deep. *House and photograph from the collection of Ray and Gail Carey.*

Although it is known that Wagner also made dollhouse furniture, very few pieces have been positively identified as such. One set is shown which includes a table, desk, two chairs, and a shelf. A label on the back of the shelf reads "D.H. Wagner & Sohn, Spielwaarenhandlung, Leipzig, Grimmaische-Strasse 6., Naschmarkt-Gegenuber." The wood furniture is finished in red and the chairs are upholstered in cream colored corduroy. The desk, chair back, and shelf have an impressed design similar to the furniture known as "Golden Oak."

Many other dollhouses, although clearly marked "Germany," are more difficult to identify as having been made by a specific company. Some may have been made by one of the companies previously discussed but are simply unusual variations or later models. It is also likely that several other German toy companies manufactured dollhouses, perhaps making only a few models for a limited time, and that records of these companies have not been found.

Three examples are shown of cottages which have been previously attributed to Gottschalk, but may have been made by Wagner or somehow related to the Wagner firm. The three houses, with their distinctive trapezoidal shaped dormers, vary slightly in size and have subtle variations which provide some clues to their origin. All three are marked "Germany" on the bottom and all show the mottled stenciling pattern, characteristic of Wagner, on their bases or porch rails. One has a door with the three vertical lines typical of Wagners and another has two lithographed windows which appear to be similar to those found on known Wagner dollhouses. A two-story version, with the same trapezoidal dormer, lithographed windows, and stenciled base is also shown. *International Dolls House News* has featured two such houses and noted that they were labeled "Moko." According to Marion Osborne, in her book *Dollhouses A-Z, 1914 to 1941*, this was a brand name used by J. Kohnstam Ltd. This company had a London address but also had facilities in Germany, including one in Sonneberg, as did Wagner. An identical two-story house is pictured in *The Universal Toy Catalog, 1924-26*, on the pages of unidentified manufacturer number 3. The *Catalog* pictures other wooden toys, including stables, trains, dollhouse furniture, arks, kitchens, and farms as being made by the same manufacturer. The door of the two-story house is divided into two panels with a lion's head door knocker. The door knocker may provide a tentative connection to another two-story house with the same feature. This house, like the previous four, has a red roof. Although this house also has some Gottschalk characteristics, the number 3073 penciled on its underside would be illogical for that series. The interior, unlike most Gottschalk houses, is plain.

Another series of one room cottages, with bright red roofs, green trim, and embossed stone bases, bears a distinct resemblance to early Schoenhut houses. However, underneath each of the houses is clearly stamped "Made in Germany" in purple ink. The cream colored example is slightly larger than the orange one. Both have identical doors and are side-opening. Both houses have lithographed decals for windows, surrounded by rather elaborate cardboard frames. The decals themselves are very similar to those found on some Gottschalk "Red Roof" houses and on some early houses by Schoenhut. The embossed stone bases are also like those on early Schoenhuts. Inside both of the cottages are plain, with no floor paper or wallpaper.

This small house shows the typical Wagner stenciling on its roof. Its front door bears a lion's head knocker ($400-600). 13.75" high x 9.25" wide x 7" deep. *House and photograph from the collection of Marilyn Pittman.*

Few pieces of furniture have been positively identified as Wagner. However, the shelf of this set bears a label which reads "D.H. Wagner & Sohn, Spielwaarenhandlung" (loosely, 'toy business') and the address. The wood furniture has a red finish and the two chairs have corduroy upholstery. The desk, chair, and table have impressed designs (Shelf $125-150, other pieces $65-100 each. The label on the shelf would increase its value and the value of the set as a whole). *Furniture and photograph from the collection of Linda Hanlon.*

There are two rooms inside the small Wagner. The lower level still has the original wallpaper. *House and photograph from the collection of Marilyn Pittman.*

German Dollhouses and Furniture 105

Three cottages of unknown origin exhibit some features typical of Wagner. The house on the left has lithographed windows similar to those often found on Wagner houses. Each house contains only one room and is marked "Made in Germany" on the bottom ($350-400 each). The houses vary slightly in size, but are approximately 8" high x 9" wide x 5.5" deep. *Houses and photograph from the collection of Ruth Petros.*

A two-story house has the same distinctive trapezoidal shaped dormer found on the three cottages. It features lithographed windows, a lion's head door knocker, and stenciled stone base, all characteristics of Wagner houses. Similar examples have been found marked "Moko" and the same house was advertised in *The Universal Toy Catalog 1924-26*. The house contains two, unpapered rooms ($650-800). *House and photograph from the collection of Linda Boltrek.*

This small house, with its red roof, could possibly have been made by Gottschalk. However, underneath the house is stamped "Made in Germany" in purple ink and the numbers "3073/58/2" are penciled. That number seems unlikely to be one of the "Red Roof" series. The paneled door, with lion's head decoration, is very similar to that found on the two story house with trapezoidal dormer ($650-800). 11.5" high x 8.25" wide x 4.5" deep. *House and photograph from the collection of Lois L. Freeman.*

The interior is plain except for its lithographed fireplace. *House and photograph from the collection of Lois L. Freeman.*

106 German Dollhouses and Furniture

One of another series of one-room cottages with red roofs, this orange model has an embossed faux stone base similar to those found on early Schoenhuts. It has elaborately framed lithographed windows and contains one room. Its base is stamped "Made in Germany" in purple ink ($350-400). 7.5" high x 9" wide x 9" deep. *House and photograph from collection of Nancy Roeder.*

A similar cottage is painted a creamy yellow with the same green trim and red roof. Like the earlier example, this one contains one unpapered room. The walls, floor, and roof are made of a heavy cardboard, but the base and corner pieces are wood. Underneath, it is stamped "Made in Germany" in purple ink.($350-400). 9.5" high x 9.75" wide x 9.75" deep. *House and photograph from collection of Marilyn Pittman.*

GERMAN WOOD DOLLHOUSE FURNITURE

Throughout most of the history of commercially produced dollhouse furniture, Germany dominated the market. From the early 1800s until World War II, the Germans were able to produce furniture of such high quality and reasonable price that British and American toy companies scarcely bothered to compete. And when they did, they often seemed to rely heavily on German ideas and designs. For this reason, often the most historically accurate furniture for a turn of the century dollhouse, even one made in England or America, would be German.

Identifying German furniture is a confusing task for collectors. Little is known about manufacturers from the early 1800s and almost all of the furniture from that time remains unattributed. It was not until later in the century that dollhouses and furniture began to be produced in quantity and at prices that made them accessible to middle class children rather than upper class adults. Catalogs and boxed sets from this time period have allowed researchers to begin identifying specific manufacturers. However this has all been complicated by a lack of a standardized vocabulary. When the furniture has been pictured in books on dollhouses and their furnishings, a variety of often contradictory, and sometimes inaccurate, names have been used in an effort to somehow categorize them. German furniture has variously been described according to the supposed place of manufacture (e.g. Waltershausen), the material used (e.g. "yellow cherry," "rosewood," or "pressed cardboard"), the finish (e.g. "red lacquer"), the style (e.g. "Beidermeier" or "dolls' house Duncan Phyfe"), or the company (e.g. "Gebruber Schneegass.") Quite often the same piece of furniture could fall under any of these categories. Some of the labels used most often by American collectors will be discussed with an attempt to include other commonly used names.

Inlaid Wood

From the early part of the 19th century, wood dollhouse furniture was produced in the rural region of Thuringia and neighboring areas. The furniture was usually simple in style, with straight lines inspired by full-scale, early Empire designs, unpainted and embellished with an inlay of contrasting wood. Similar furniture, from the Saxony region, was often painted red and trimmed in black leather.

Faux Grained Wood

In the mid-1800s furniture was produced with a wood graining which provided an inexpensive means of imitating walnut. The decoration gave the furniture a rich look when applied to shapes that were still rather simple. To modern collectors, in the age of Formica, the furniture can have the misleading appearance of cheap veneer. The furniture had working doors and drawers. Some pieces were upholstered and some had turned legs. Such furniture has been found in the circa 1850 catalog of manufacturer Johann Daniel **Kestner**, Jr. of Waltershausen. However, this furniture was made over a long period of time and with a wide variation in both size and quality, so it seems likely that other manufacturers may have also used this technique.

Carved Furniture from the Black Forest

This very rare furniture was made in 1" and larger scales circa mid-1800s. It was highly detailed with many elaborately carved parts. Although the intricacy of the craftsmanship might lead one to believe that the pieces were handmade, the existence of the original box proves that it was commercially manufactured. The box contains a lot of information but, unfortunately, does not give the name of the company .

Biedermeier

Vivien Greene has identified **Gebruder Schneegas und Sohne** (the brothers Schneegas, and later, son) as the manufacturer of the furniture most commonly known to American collectors as Biedermeier. The furniture was produced in Waltershausen, a small town in the state of Thuringia, circa 1830-45 and on through the early twentieth century. The word "Biedermeier" refers to a style of full size furniture popular in Europe during the early 1800s. The "real" furniture used light-colored woods and had a simplicity of form which was said to be a reaction to the highly ornamented French Empire style. The furniture experienced a popular revival in the late 19th century (when most of the dollhouse furniture was made) and at this time the term "Biedermeier" was first used in a derogatory way to denote the style as old-fashioned. It is unclear exactly how this came to be the name most frequently used to describe the dollhouse furniture characterized by imitation ebony and rosewood finishes embellished with gilt transfers. The furniture has also been referred to as **"Waltershausen"** in honor of its place of manufacture or **"dolls' house Duncan Phyfe"** after the American furniture maker. Some of the pieces had marble tops, glass doors, and pewter knobs. Sofas and chairs were usually upholstered in silk or printed cotton. It was made in a range of sizes from approximately 3/4" to the foot to a very large 1" to the foot scale. The styles ranged from rather simple lines to Rococo and, more rarely, Gothic Revival. Much of the furniture could loosely be described as in the Regency style. Similar furniture was also made with lighter, even white, finishes, but these pieces are very rare. Not all of the Biedermeier furniture had the gilt transfers. Some of the pieces were handpainted and others were plain. The transfers themselves were printed in hundreds of different patterns, some of which depicted historic sites or scenes. The various styles are believed to have been in production simultaneously up until after World War I and therefore can not be used to accurately determine the age of a particular piece. The quality of construction and the degree of ornamentation varied widely, so it is possible that similar furniture may have been produced by other companies in the same area.

Golden Oak

Schneegas also made furniture in a variety of lighter woods and finishes which gave rise to many different, often confusing names. This furniture, which resembled the popular American mass-produced furniture known as **Golden Oak**, became known to American collectors by that name. Most of it was actually made of **yellow cherry**, the more accurate term used for the furniture in Europe. The furniture was most often finished in a honey color, but darker brown or red stains and lacquers were also used. Depending on the type of wood or finish, it has been variously labeled **Fruitwood, Honey Maple, Red Cherry, Red Lacquer,** and **Red Mahogany**. The furniture is also called **"Waltershausen,"** because it was manufactured in that town, although this name seems to be applied more often to the Biedermeier line. The Golden Oak furniture was produced in at least three scales, ranging from approximately 3/4" to 1 1/2" to the foot and came in many styles from Gothic Revival to Arts and Crafts. Like the Beidermeier furniture, many of the pieces had marble tops, turned legs, pewter knobs, and glass doors. Dining room chairs sometimes had cane seats and some pieces were upholstered in velvet or cotton, often embellished with fringe. Some items, especially beds and cabinets, had a decorative cornice with one to three which had an Art Nouveau flourish. At least one other company, D.H. Wagner, was known to have made furniture with impressed designs during this same period. Much of the Schneegas furniture was exported to the United States so it is not uncommon to find pieces with labels from F.A.O. Schwarz or other stores.

Red Line

A wide variety of wood pieces known to collectors as **"Red Line"** or **"Red-stained"** were produced in Germany from the beginning of the twentieth century to the 1930s. The name was derived from the reddish stain used to finish the furniture. Many of the pieces were marked "Germany" either incised, stamped, or on a paper label. Most were manufactured in 1" to the foot scale, but some pieces were slightly larger and some as small as 3/4" to the foot in scale. The furniture was quite detailed, with operable drawers and doors, turned legs, and glass or isinglass panels. Some pieces were upholstered in prints or velvet. The furniture was made in many different styles. Some of the early pieces may have had impressed designs, but they were more commonly made in the modern and revival styles popular circa 1920s. It is not possible to attribute this furniture to any one company. Schneegas used a reddish stain on some of their furniture and it is possible that some of it would fall into this category. Collectors should keep in mind that full-size mahogany furniture was popular during this period and it may be assumed that different dollhouse furniture manufacturers would reflect that in their products.

Lithographed Paper

Once believed to have been French in origin, it is now generally accepted that dollhouse furniture made of **lithographed paper** over wood or cardboard was really produced in Germany. Using the same process as on lithographed dollhouses, the elaborately printed paper allowed manufacturers to provide a high level of ornamentation in an inexpensive way. The brightly colored lithography, sometimes referred to as **chromolithography**, was often embellished in gold. The potential subject matter was limitless and included faux wood grains, flowers, cherubs, animals, children, and even insects. Some of the early, circa 1870, furniture had turned legs and velvet upholstery which, when combined with extravagant lithography, was very impressive. At the other extreme, designs were lithographed onto paper or cardboard and used on furniture made of pieces of cheap, flat wood. The complexity possible with lithography made even these pieces desirable.

Having dispensed with the most obvious categories, the collector is still left with a tremendous amount of wood furniture, clearly marked "Germany" but with no other known attribution. It is included here, divided loosely by similar characteristics or intended function, in the hope that future researchers will be able to provide additional information.

Wood furniture with simple lines, embellished with inlaid designs of contrasting woods, were produced in Thuringia and neighboring areas from the early part of the 19th century. The furniture is approximately 3/4" to the foot in scale (Chairs $35 each, sofa $45-50, tables $40-45 each, chest $45-50). *Furniture from the collection of Dian Zillner. Photograph by Suzanne Silverthorn.*

On this 3/4" to the foot scale bed from the same region, the contrasting wood decoration is applied to the surface rather than inlaid ($65-75). *Furniture from the collection of Dian Zillner. Photograph by Suzanne Silverthorn.*

Faux grained furniture was produced by a variety of companies including Johann Daniel Kestner, Jr. circa 1850. This later set, in a large 3/4" to the foot scale, includes pieces upholstered in velvet with gilt paper trim (Sofa $65-75, table $40-45, pier mirror $45-50, cabinets $60-70 each, chairs $35 each). *Furniture and photograph from the collection of Patty Cooper.*

These chairs are typical of the red-painted, leather trimmed furniture produced in the Saxony region of Germany in the early part of the 19th century and onward. They are in approximately 1.25" to the foot scale ($40-50 each). *Furniture from the collection of Dian Zillner. Photograph by Suzanne Silverthorn.*

Another faux grained parlor set was made with a lighter finish (Sofa $65-75, table $40-45, mirror $45-50, tables $40-45 each, cabinet $60-70, chairs $35 each). *Furniture and photograph from the collection of Sharon Unger.*

The existence of this boxed set verifies the commercial origins of the carved furniture from the Black Forest area of Germany (Not enough examples to determine price). *Boxed set from the collection of Paige Thornton. Photograph by Steve Thornton.*

German Dollhouses and Furniture 109

This elaborately carved, very rare furniture was made in 1" to the foot and larger scales (Not enough examples to determine price). *Furniture from the collection of Paige Thornton. Photograph by Steve Thornton.*

The upright piano and fireplace are especially rare and desirable to collectors. The sofa and chairs are upholstered in velvet (Not enough examples to determine price). *Furniture from the collection of Paige Thornton. Photograph by Steve Thornton.*

Biedermeier furniture was produced by Gebruder Schneegas in Waltershausen from the early 1800s through the early 1900s. This very old, very rare Beidermeier set is in the Gothic style (Not enough examples to determine price). *Furniture from the collection of Paige Thornton. Photograph by Steve Thornton.*

Gilt transfers were used to embellish the imitation rosewood furniture known as Beidermeier. The process allowed Schneegas to produce highly ornamented furniture so economically that British and American manufacturers scarcely bothered to compete. The items shown are in 1" to the foot scale (Sofa $175-200, china cabinet $275-350, table $125-150, chairs $45-50 each). *Furniture and photograph from the collection of Ray and Gail Carey.*

Three other very old Beidermeier pieces are shown, including a square drop-leaf table and a writing desk with chair (Not enough examples to determine price). *Furniture from the collection of Paige Thornton. Photograph by Steve Thornton.*

110 German Dollhouses and Furniture

A 1" to the foot scale Biedermeier bedroom set shows the rich detail made possible by the use of gilt transfers. Note that the construction of the dressing table is actually rather simple with no turned pieces or operable parts, but the overall effect is striking (Small chest $100-125, other pieces $150-185 each). *Furniture and photograph from the collection of Ray and Gail Carey.*

A two-piece Beidermeier bedroom set is shown. The mirrored chest has a marble top and two opening drawers. Both are in 3/4" to the foot scale (Chest $135, bed $165). *Furniture from the collection of Dian Zillner. Photograph by Suzanne Silverthorn.*

Beidermeier parlor sets were also made in 3/4" to the foot scale ($300-400) *Furniture from the collection of Dian Zillner. Photograph by Suzanne Silverthorn.*

A 1" scale Beidermeier writing desk has opening doors, letter compartments, and a marble writing surface. The non-opening drawers are suggested by gilt transfer outlines ($350-375). *Desk and photograph from the collection of Ray and Gail Carey.*

The early Schneegas furniture, known as Golden Oak or Yellow Cherry, featured marble tops, turned legs, and cornices decorated with three holes. The bed is 7" long and the other pieces are in a comparable large 1" to the foot scale (Chair or hanging shelf $50-65 each, sewing table $110-125, larger pieces $150-250 each). *Furniture and photograph from the collection of Ray and Gail Carey.*

German Dollhouses and Furniture 111

Golden Oak pieces were made in scales to fit even small dollhouses. The simply-made marble top dresser ($35) and chair ($25) are in a small 3/4" to the foot scale. The beds ($45 each) were probably intended as youth beds to accompany 1" scale furniture. The cabinet with impressed design ($45) can be used as a small scale wardrobe but could also be a 1" scale medicine cabinet. *Furniture and photograph from the collection of Patty Cooper.*

The Schneegas furniture was also made in somewhat simpler versions. In this grouping, the mirrored chest does not have a marble top and the headboard and footboard each contain only one hole ($100-150 each). *Furniture from the collection of Dian Zillner. Photograph by Suzanne Silverthorn.*

Three examples are decorated with the impressed designs often used on the Schneegas furniture known as Golden Oak. The piano has three depressible keys ($100-125). The clock is 8.5" tall ($125). The wardrobe is somewhat smaller in scale and has two opening doors ($85-100). *Furniture and photograph from the collection of Patty Cooper.*

A Golden Oak desk has a slant top which can be lifted up and drawers on the side ($85-125). The piano has three depressible keys ($100-125). *Furniture from the collection of Dian Zillner. Photograph by Suzanne Silverthorn.*

Many of the same impressed designs were used for Schneegas furniture with a Red Lacquer finish, such as these pieces in 1" to the foot scale (Chairs $65 each, table $100, nightstand $125, bed or three drawer chest $175-200 each). *Furniture and photograph from the collection of Ray and Gail Carey.*

Impressed designs were also used to embellish Golden Oak furniture in 3/4" to the foot scale (Chairs $20-25 each, bed $65, three drawer chest $60, mirror $25, washstand $50-60, mirrored wardrobe $65-75). *Furniture and photograph from the collection of Patty Cooper.*

These Schneegas Red Lacquer pieces, in large 1" to the foot scale, are rare. The overstuffed chair is covered in maroon leatherette ($100) as is the chaise with impressed designs along the backbar ($375). The grandfather clock has weights and a pendulum in a recessed compartment ($350). The kneehole desk has a writing surface protected by green felt, three operable drawers and two doors, all with impressed designs. The matching desk chair has impressed designs and a green felt seat (Two piece set $395). The remaining chair has a slat back, with turned front legs and a seat of woven straw ($95). *Furniture from the collection of Paige Thornton. Photograph by Steve Thornton.*

The Schneegas dining room table in Red Lacquer pulls apart to hold two leaves. The center section is stabilized with a fifth leg ($350). The four matching chairs have turned front legs, impressed designs on their backs, and woven straw seats (Set of four chairs $400). *Furniture from the collection of Paige Thornton. Photograph by Steve Thornton.*

German Dollhouses and Furniture 113

Red-stained German bedroom pieces in 1" to one foot scale include a blanket chest ($40), tall mirrored chest of drawers ($60), straight chair ($25), four poster bed ($50), two drawer sewing table ($55), and upholstered chaise lounge ($60). *Furniture and photograph from the collection of Ray and Gail Carey.*

The maker of this Red-stained German furniture is not known. Most of the pieces were marked "Made in Germany." The 1" to one foot scale living room pieces include an upholstered rocking chair ($45), wing chair ($40), and sofa ($60). The fireplace has brick paper and a built-in grate ($35). The mirrored sideboard has two opening drawers and doors ($55). The upholstered shield-backed chairs ($35 each) and double-pedestal table ($40) reflect styles popular in the 1920s and 1930s. *Furniture and photograph from the collection of Ray and Gail Carey.*

Some matching Red-Stained dining room pieces include a bow-front china cabinet ($75), two Windsor side chairs ($25 each), a drop-leaf table ($65), and a wheeled tea-cart ($65). *Furniture from the collection of Dian Zillner. Photograph by Suzanne Silverthorn.*

A similar four poster bed ($50) is accompanied by pieces in a different style including a Windsor chair ($40), dressing table ($65), and nested tables ($65). *Furniture from the collection of Dian Zillner. Photograph by Suzanne Silverthorn.*

Other 1" to one foot scale pieces of the German Red-stained furniture are shown: A server with two opening doors ($45), another set of three nested tables ($65), an unusual bed with fan-shaped head and footboards ($60), a trefoil-shaped, tilt-top table ($50), and a gateleg dining table ($70). *Furniture and photograph from the collection of Ray and Gail Carey.*

This grouping of Red-stained German furniture includes some pieces that would be suitable for a library. The lowboy has four drawers ($40). The glass-fronted cabinet contains faux books ($70). The magazine rack ($15) is quite contemporary in design while the sewing chest ($55) is in an earlier style. The library table has a beveled edge and two turned legs joined by a stretcher ($35). *Furniture from the collection of Dian Zillner. Photograph by Suzanne Silverthorn.*

Red-stained furniture, with paper labels marked "Germany," was made in 3/4" to one foot scale. It is not known whether these pieces were made by the same company as the 1" scale furniture. The dining room chairs have turned front legs and the buffet has an opening door decorated with an impressed design (Set $85-100). *Furniture and photograph from the collection of Patty Cooper.*

Designs were also lithographed onto cardboard and combined with simple shapes cut out of flat wood to create attractive pieces in a very economical way (Set $50-75). *Furniture and photograph from the collection of Patty Cooper.*

German Dollhouses and Furniture 115

A lavishly decorated set of 1" to one foot scale furniture, with a geometric pattern, illustrates the richness of design made possible by lithography. The set in some ways resembles the Biedermeier furniture in that the designs appear to be more like transfers or decals than solid pieces of lithographed paper. The washstand ($110) and night table ($95) have marble tops and opening doors. The pier mirror ($110) also contains a drawer. The bed has turned feet ($150) and the table ($110) has a rather complex pedestal base. The sofa ($125-150) and two chairs ($50-60 each) have turned legs and are upholstered in a printed fabric with fringe. *Furniture from the collection of Madeline Large. Photograph by Patty Cooper.*

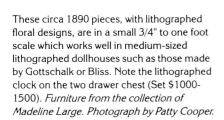

These circa 1890 pieces, with lithographed floral designs, are in a small 3/4" to one foot scale which works well in medium-sized lithographed dollhouses such as those made by Gottschalk or Bliss. Note the lithographed clock on the two drawer chest (Set $1000-1500). *Furniture from the collection of Madeline Large. Photograph by Patty Cooper.*

The 3/4" to the foot set is embellished with pictures of birds and flowers. The chairs have turned legs and upholstered velvet seats. The cabinets have opening doors and one has a lithographed clock (Set $700-900). *Furniture from the collection of Dian Zillner. Photograph by Suzanne Silverthorn.*

116 German Dollhouses and Furniture

Another cabinet with a lithographed clock has an almost bizarre design of insects on flowers. Imagine a two feet long beetle decorating the sideboard in a full-sized dining room! ($75-85). *Furniture and photograph from the collection of Sharon Bernard.*

These 1" to the foot scale green dining room pieces are stamped "Made in Germany" inside a rectangle ($10-15 each). *Furniture from the collection of Dian Zillner. Photograph by Suzanne Silverthorn.*

Although somewhat smaller, this four-piece set in 3/4" to the foot scale has the same rectangular "Made in Germany" stamp and the same type seat cushion as the green chairs in the previously shown dining room set ($20-25 each). *Furniture from the collection of Dian Zillner. Photograph by Suzanne Silverthorn.*

These parlor pieces, in a small 1" to the foot scale, have an oval stamp with the words "Made in Germany," but the upholstery is very similar to that used on the green pieces with the rectangular stamp (Set $75). *Furniture from the collection of Dian Zillner. Photograph by Suzanne Silverthorn.*

Bedroom pieces in the same large 3/4" to the foot scale are also stamped with "Made in Germany" inside a rectangle ($20-25 each). *Furniture from the collection of Dian Zillner. Photograph by Suzanne Silverthorn.*

German Dollhouses and Furniture 117

Another set, in a small 1" to the foot scale, has the same oval "Made in Germany" stamp (Set $75). *Furniture from the collection of Dian Zillner. Photograph by Suzanne Silverthorn.*

Upholstered in a textured fabric, these 1" scale parlor pieces are believed to be German, but the manufacturer is unknown (Set $250-300). *Furniture and photograph from the collection of Ray and Gail Carey.*

Parlor pieces covered in velvet were made in the 3/4" to the foot scale appropriate for medium-sized Bliss or Gottschalk houses. The unmarked pieces are believed to be German. They have turned legs and are trimmed with fringed braid (Set $175). *Furniture and photograph from the collection of Patty Cooper.*

This tiny grouping, in approximately 1/2" to the foot scale, contains pieces which are upholstered in velvet and have turned legs. The circa 1900 chairs and sofa, with bulbous legs, are trimmed in gilt paper (Set $175). The piano, which is not as old, has a hinged lid and is marked "Kranich & Bach" above the keyboard ($50-65). *Furniture and photograph from the collection of Patty Cooper.*

Some of the German furniture, imported into the United States for many decades, was very inexpensively made. These upholstered parlor pieces, in 3/4" to the foot scale, are cut out of flat pieces of cheap wood and have serrated paper fringe (Set $50). *Furniture and photograph from the collection of Patty Cooper.*

118 German Dollhouses and Furniture

A pedestal dining table with two chairs and matching china cabinet, circa 1920s, are stamped "Made in Germany" and also bear a paper label from F.A.O. Schwarz (Set $150). *Furniture and photograph from the collection of Ray and Gail Carey.*

Nursery furniture can be somewhat difficult to find because few American companies made such items. The pieces shown here are 3/4" to the foot in scale and are stamped "Germany." (Set $150). *Furniture and photograph from the collection of Patty Cooper.*

The bedroom set shown is part of a group of 1" to the foot scale furniture which is clearly marked "Germany" but includes pieces which are very similar to the 1928 Schoenhut line. The mirrored vanity and chair with cut-out back are distinguishable from their Schoenhut counterparts only through such minor details as the mitering of the mirror or the turning of the front legs ($100). *Furniture and photograph from the collection of Patty Cooper.*

Some of the pieces from this circa 1920s German bedroom set are similar to ones sold by Tynietoy, especially the rush-seated rocking chair (Set $200-225). *Furniture and photograph from the collection of Ray and Gail Carey.*

German Dollhouses and Furniture 119

Collectors are delighted to find sets as complete as this 1" scale German bedroom set ($200-250). *Furniture and photograph from the collection of Ray and Gail Carey.*

Bedroom sets of this type have been seen in several variations, including white enamel and mahogany finishes. The beds have attached mattresses and are approximately 5" long. They could be considered to be 3/4" to the foot in scale or as youth beds in a small 1" to the foot scale. Complete sets often include chairs which are fully 1" to the foot in scale. Circa 1910-1920 (Beds $35 each, armoire $45, pier mirror $40, nightstand $15). *Furniture and photograph from the collection of Patty Cooper.*

Simply cut from flat-pieces of inexpensive wood, these large 3/4" scale German pieces give the impression of having turned legs (Bed $15, nightstand with attached lamp $12, other pieces $15 each). *Furniture and photograph from the collection of Patty Cooper.*

Kitchen pieces like these have been found in the same lots as the furniture shown in the previous photo. They are quite similar in scale, construction, and quality, and may have been made by the same company. Most of the pieces are stamped "Germany." ($10-20 each). *Furniture from the collection of Dian Zillner. Photograph by Suzanne Silverthorn.*

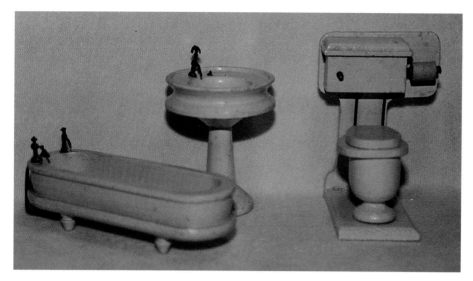

Overscale 1" to the foot bathroom pieces like these seem meant to celebrate the innovation of indoor plumbing. They were equipped with wire pipes, elaborately molded pewter handles and faucets, and even toilet paper. A nearly identical set was advertised in the 1921 Sears, Roebuck and Co. catalog for 69 cents. Similar toilets have been found with high tanks and pullchains (Set $150). *Furniture and photograph from the collection of Ray and Gail Carey.*

German bathroom sets were also made in various ceramic materials. This set has metal faucets and a wooden seat. It is 1" to the foot in scale and is very similar to bathroom pieces sold by the Wisconsin Toy Company in the 1920s (Set $90-100). *Furniture from the collection of Dian Zillner. Photograph by Suzanne Silverthorn.*

GERMAN METAL DOLLHOUSE FURNITURE

Rock and Graner

Rock and Graner of Wurttemberg was established in 1813 and manufactured many types of metal toys. Sometime in the mid-1800s, they began producing pressed tinplate dollhouse furniture with faux painted surfaces which resembled wood or, sometimes, upholstery, although more often the pieces had real silk or velvet upholstery. The furniture was often embellished with pierced metal fretwork and some of the serpentine or cabriole legs were in cast, rather than pressed metal. Most of the furniture was lacquered in a shade of brown, but white or ivory was used for kitchen and bathroom items. Each piece was handpainted and often decorated with little gold beads or other trim. The high quality craftsmanship made the furniture expensive to manufacture and contributed to the company's inability to compete with the more modern methods being used by other companies at the beginning of the twentieth century. Rock and Graner furniture was made in a range of sizes and styles. Most of the furniture could be described as rococo and was in a large 1" to the foot scale, which due to the lightweight nature of the materials seems less imposing than the wood pieces in the same size. The firm went out of business around 1904. Furniture by Rock and Graner commands a very high price among today's collectors.

Rock and Graner furniture was made of lacquered tin from the mid-1800s through 1904. These exquisitely crafted pieces are in great demand by collectors. This grouping, in a large 1" to the foot scale, includes a youth bed, two wardrobes, a cabinet with pierced gallery, a draped bed, and a marble-topped nightstand (Not enough examples to determine price). *Furniture from the collection of Paige Thornton. Photograph by Steve Thornton.*

A Rock and Graner parlor set is shown with hand-applied wood-graining and cast metal cabriole legs (Not enough examples to determine price). *Furniture and photograph from the collection of Anne B. Timpson.*

Marklin

Like many German toy companies, Marklin was located in the Wurttemberg area. Established in 1859 by tinsmith Theodor Friedrich Wilhelm Marklin, the company was originally known as W. Marklin. Caroline Marklin, Theodor's wife, was very active in the business and is said to have been one of the first female sales representatives. She is responsible for keeping the business afloat after her husband's untimely death in 1866. Their sons eventually took over the business and in 1888 the name was changed to Gerbruder Marklin (for the brothers Marklin.) Often called the king of German toymakers, their best known products included toy metal cars, boats, airplanes, trains, and train stations. However, some of the first items produced by the company were metal pieces for dollhouse kitchens. By 1895, the company was offering a wide range of lacquered tinplate dollhouse furniture, including stoves, fireplaces, bathroom furnishings, beds, and accessories, usually in 1" to the foot scale or slightly larger. The Universal Toy Catalog of 1924-26 designates Marklin as manufacturer number 14 and shows several of their stoves.

Marklin is one of the few German toy manufacturers to have survived both world wars. By the mid-1930s, Marklin began to concentrate almost exclusively on their model railroad products. Although the company was involved in some war-related production during World War II, the plant escaped any direct damage. Soon after the war ended, they started producing model trains again, this time with an increasing number of plastic parts. The Marklin company has been in continuous production since its establishment and today is largely known for its production of HO scale railroads. Marklin catalogs have been reprinted which allow collectors to verify pieces made by that company with some accuracy. The Marklin name carries a certain cachet, accompanied by high price, which contributes to the erroneous attribution of many pieces to Marklin. Other German companies, such as Bing, are known to have made tin furniture during this same time period.

This large 1" to one foot scale bed is one of the more commonly found pieces by Marklin. It is highly detailed with a curlicued headboard and real springs supporting the mattress ($250). *Bed from the collection of Madeline Large. Photograph by Patty Cooper.*

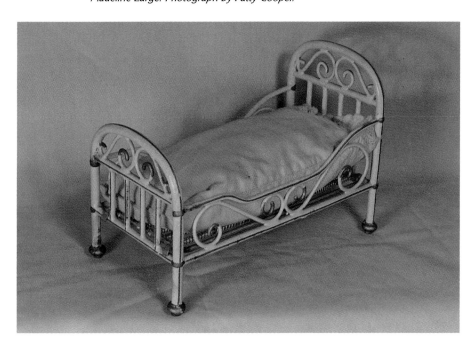

The lamp on the right is a rare piece made by Marklin ($250-300+). The rare lithophane is believed to have been made by Babette Schweitzer ($500+). *Accessories and photograph from the collection of Anne B. Timpson.*

A rare Marklin pier mirror, circa 1910, is illuminated by an electrified bulb (Not enough examples to determine price). *Mirror and photograph from the collection of Anne B. Timpson.*

Soft Metal

Other highly collectible German furniture and accessories were made of soft metals such as cast pewter. Many of these items were first made in the late-1700s and at least one such company, **Babette Schweitzer** of Diessen, is still in business. The Schweitzer firm made a wide variety of accessories and furniture, in a variety of scales, molding pewter into elaborate filigreed designs. Another leading manufacturer of soft metal dollhouse items was **F.W. Gerlach** of Nuremberg, known primarily for their high quality, detailed accessories, many of which were sold by Tynietoy. Gerlach was in business from the mid-1800s through 1939. (See the Accessories chapter for many examples of soft metal accessories made in Germany.)

Typical of pieces produced by the Deissen firm of Babette Schweitzer, this bedroom set is made of elaborate filigreed metal ($100-150 each). *Furniture and photograph from the collection of Anne B. Timpson.*

German Dollhouses and Furniture 123

The small bed, a little over 1/2" to the foot in scale, is believed to be a later product of Babette Schweitzer. The fireplace may be from the same company ($45-60 each). *Furniture and photograph from the collection of Patty Cooper.*

These circa 1900-1910 German soft-metal pieces are in approximately 3/4" to the foot scale, an ideal size for many of the lithographed houses by Gottschalk or Bliss. The chaise ($50-65) has silk upholstery while the sofa ($100-110) is upholstered in velvet. The desk ($65) has straighter, more modern lines and contains a small typewriter ($10-15). The floor lamp ($35-45) has a fringed shade and a socket for a small lightbulb. The gold curio cabinet ($65) has an isinglass panel in its door. The upright piano is missing its keyboard cover but still has two adjustable candleholders and its original bench (Piano with stool $150-195). The marble topped fireplace is similar to the one shown in the earlier photograph and may also have been made by Schweitzer ($50-65). *Furniture and photograph from the collection of Patty Cooper.*

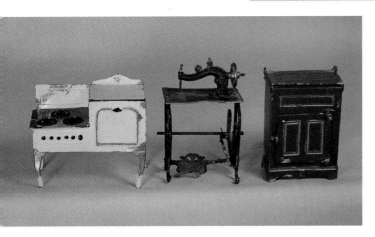

These enameled metal bathroom pieces, in 1" to the foot scale, may originally have been part of a separate German bathroom ($75-100 each). *Furniture and photograph from the collection of Patty Cooper.*

Items for domestic chores were also made in 3/4" to the foot scale soft metal, including a gas-type stove ($50), sewing machine with movable treadle ($65), and icebox with opening doors on the top and front ($50). *Furniture and photograph from the collection of Patty Cooper.*

OTHER WOOD DOLLHOUSES AND FURNITURE

HALL'S LIFETIME TOYS

Hall's Lifetime Toys was based in Chattanooga, Tennessee, for over thirty-five years. The company was begun by Charlie Hall in 1942 after he successfully introduced a dollhouse canopy bed at the Toy Fair in New York City that same year. With 2,000 orders to fill, he rented a building, secured some part time help, and became a toy manufacturer. All of the toys made by the Hall Company carried a lifetime guarantee.

Mr. Hall continued with the business until he died in 1959. Then his widow carried on the company until William Keiss and George Blackwell Smith bought into the firm in 1974. These two men, along with Mrs. Hall, continued to sell dollhouses and furnishings until the late 1970s. The firm produced many different designs of 1" to one foot scale dollhouses including several Colonial designs, farm houses, Tudor houses, and Cape Cod designs. The Hall's houses were made in either one story, two story, or three story designs. Hall's also offered historical houses including Mount Vernon, Hermitage House, Molly Pitcher, Abigail Adams, Betsy Ross, and a Williamsburg home. Most of the houses had four to nine rooms each. The row houses were made with either three or four rooms. Many of the Hall's houses were equipped with electric lights. During the 1970s, the houses ranged in price from $50 to $270.

There were several lines of wood 1" to one foot scale furniture made for the houses of the 1970s. These included the plain line which sold by the box, the Cherry Fine Line sold by the piece, and the Little White Line also sold by the piece. Many of the White Line items used the same design as The Fine Line. Nicer kitchen and bathroom sets were also offered to accompany the more expensive furniture. These included a stove, refrigerator, sink, table and chairs, tub, vanity basin, hamper, and commode. Another line was offered which was called Country Pine but it was larger in scale and was to be used with dolls 7" to 12" tall. In the earlier catalogs, the Cherry Fine Line was called the Mini-Line and although the designs were the same, the wood was not identified. The more expensive furniture included functional parts but the cheaper boxed pieces did not. The boxed furniture sold for $12 per box in 1976 while the individual pieces of the more expensive lines ranged from around $4.00 to $14.00 for each item. Hall's also included many accessories in their mail order catalog including rugs, fireplaces, lamps, and a patio set consisting of a chaise, table, and chair.

Hall's Lifetime Toys produced several lines of 1" to one foot furniture during the 1970s. Pictured are the cheaper kitchen pieces which sold for $12 in a boxed set in 1976 (Appliances $12-15 each, chair $8-10 each). *Furniture from the collection of Dian Zillner. Photograph by Suzanne Silverthorn.*

Hall's 1" scale cheaper dining room furniture featured an oversize corner cabinet ($15-18), buffet ($15-18), table ($15-18), and two chairs ($8-10 each). *Furniture from the collection of Dian Zillner. Photograph by Suzanne Silverthorn.*

Other Wood Dollhouses and Furniture 125

Hall's boxed 1" scale living room set (boxed set $75 and up) included a sofa, ($15-18), chair ($15-18), coffee table ($12-15), and a television ($15-18). *Furniture and photograph from the collection of Patty Cooper.*

The nursery pieces included a crib ($15-18), playpen ($15-18), chifforobe ($15-18), and a clothes hamper ($8-10). The drawers are not functional although the clothes hamper does open. *Furniture from the collection of Dian Zillner. Photograph by Suzanne Silverthorn.*

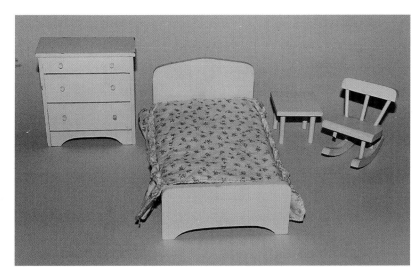

The bedroom from the 1976 boxed furniture sets included a bed ($15-18), chest of drawers ($15-18), rocking chair ($15-18), and a lamp table ($10-12). *Furniture and photograph from the collection of Lois L. Freeman.*

Hall's bathroom furniture consisted of a toilet (12-15), hamper ($8-10), bathtub ($12-15), and a lavatory ($12-15). *Furniture and photograph from the collection of Lois L. Freeman.*

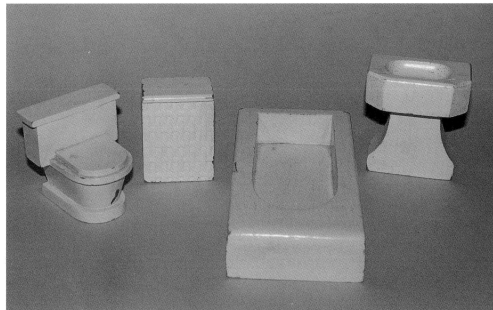

126 Other Wood Dollhouses and Furniture

The bed from Hall's Mini-Line is pictured along with two dining room items from the cheaper boxed line of furniture (Bed $35-40). Other items made for the bedroom included a cradle, wardrobe, nightstand, dresser, and chest. *Furniture and photograph from the collection of Patty Cooper.*

A 1972 catalog page from Hall's Lifetime Toys shows the more expensive "Mini-Line" furniture. The living room included a sofa, chair, rocker, and television. The dining room pieces consisted of a corner cabinet, hutch, sideboard, round table, and two chairs. The prices ranged from $4.00 to $7.50 for each item. *Catalog from the collection of Lois L. Freeman. Photograph by Suzanne Silverthorn.*

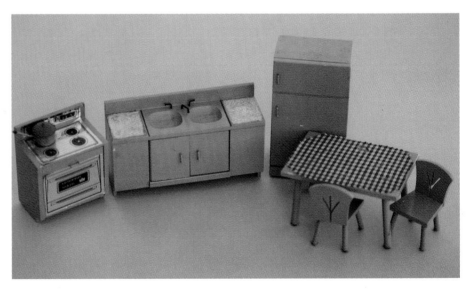

The kitchen furniture from the Mini-Line included working doors ($18-22 each). *Furniture from the collection of Dian Zillner. Photograph by Suzanne Silverthorn.*

The Mini-Line bathroom pieces also contained moving parts ($18-20 each). *Furniture from the collection of Dian Zillner. Photograph by Suzanne Silverthorn.*

Other Wood Dollhouses and Furniture 127

Hall's "Country Classic Dollhouse" dates from 1968. It is in 1" scale and was made from wood paneling. The weather vane is a replacement although it is like the original ($300-400). 21" high x 38" wide x 25" deep. *House and photograph from the collection of Lois L. Freeman.*

The inside of the "Country Classic Dollhouse" was covered with contact paper. The roof could be raised to provide more access to the upstairs rooms. The second story contained two moveable room partitions. *House and photograph from the collection of Lois L. Freeman.*

Hall's "Cape Cod Manor House" dating from 1972. Constructed of wood and hardboard, it is also in the 1" to one foot scale. In 1972 the house was priced at $80. By 1976 the cost had increased to $140 ($500-600). 27" high x 39" wide x 15" deep. *House and photograph from the collection of Lois L. Freeman.*

The Cape Cod house contained six rooms which included a fireplace, mirror, bookcases, and a carpeted stairway. The house still includes its original wallpaper. *House and photograph from the collection of Lois L. Freeman.*

Hall's "Victorian House" from 1974 which contains six rooms. The house included electric lights, real wallpaper, carpeted stairway, and simulated lap siding in front ($500-600). 26.5" high x 33" wide x 15" deep. *House and photograph from the collection of Lois L. Freeman.*

Hall's "Betsy Ross" house dating from 1976. This house represented a famous Row House from Colonial times. In 1976 it was priced at $50 ($300-400). 30" high x 14" wide x 12" deep. *House and photograph from the collection of Lois L. Freeman.*

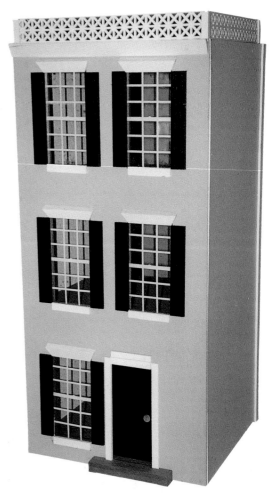

Other Wood Dollhouses and Furniture 129

The interior of the "Betsy Ross" house contained a removable staircase that was sold separately by Hall's. The three rooms are decorated in white. *House and photograph from the collection of Lois L. Freeman.*

Hall's "Molly Pitcher" row house is very similar to the "Betsy Ross" model. The house has a wood front, masonite type sides, and contains three rooms. These houses were probably especially popular during the United States Bi-Centennial year in 1976. *House and photograph from the collection of Lois L. Freeman.*

LYNNFIELD-BLOCK HOUSE-SONIA MESSER

It is hard to believe a dollhouse furniture line could continue to be sold for over fifty years. That is the case with the 1" to one foot furniture known as Lynnfield, then H. M. Miniatures, later Andi Imports, and finally Sonia Messer. Throughout most of these years, the furniture was marketed by Block House, Inc.

This furniture remained popular for so long because most of the designs were not contemporary and never went out of style. The furniture was first developed by Chester H. Waite in the early 1930s. Although he lived in Lynnfield, Massachusetts, most of the furniture was sold through Block House, Inc.

Block House, Inc., founded by Paul Frick, was an import and wholesale company located in New York City. The company also served as a

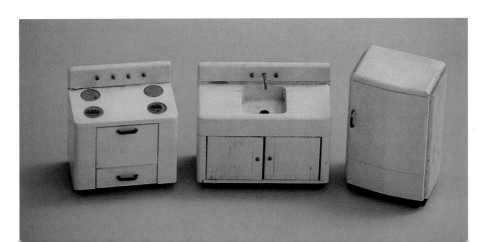

Lynnfield wood kitchen furniture in the 1" to one foot scale sold in 1938. The doors and drawers are functional. Pictured are the stove, sink, and refrigerator ($30-40 each). *Furniture from the collection of Dian Zillner. Photograph by Suzanne Silverthorn.*

representative for the makers of toys and miniatures. The Lynnfield-Block furniture was also carried by large department stores such as F.A.O. Schwarz and Marshall Fields for decades.

Included in the early line of furniture (circa 1940) was a Duncan Phyfe dining room, modern "blonde" bedroom, enameled bedroom, Empire bedroom, enameled nursery, enameled kitchen, Early American dining room, and 18th Century living rooms.

The enameled kitchen set consisted of a stove, sink, refrigerator, table, chairs, breakfast nook, and two different cabinets. One of the cabinets was an open shelved peasant unit similar to the one offered by Tynie Toy. Several of these pieces of furniture were decorated with painted designs. The enameled nursery included a crib on wheels, potty chair, chest of drawers, and small table and chair. A hard to find set from this early period is the enameled bedroom furniture which featured the new round mirrors on both the dresser and vanity. The "blonde" bedroom design included a double bed, upholstered chair and footstool, straight chair, chest of drawers, bedside table, vanity, and stool. The pulls on these pieces were quite wide and the mirrors were large and round. This furniture was typical of the early 1940s style and was offered periodically as part of the line as late as the early 1960s. The other furniture pictured in the 1940 catalog was much more traditional. Two different pianos were sold. One was a grand piano while the other was a spinet model. Duncan Phyfe designs included a sofa, drum table, dining room set, and desk. Empire styles were similar to the pieces that continued to be sold for decades. A maple Early American dining set was offered which included a rectangular table, large breakfront, ladder-back chairs, benches, and a server. Lamps, candlesticks, pictures, screens, and a floor model radio were also part of the line circa 1940.

According to Dee Snyder, ("The Collectables," *Nutshell News*, January 1987) extensive changes were made in the furniture designs around 1944. The furniture carried by Block House, Inc. was redesigned by Frick's son-in-law, Henry Messerschmidt and he and his wife began producing most of the pieces with help from outside contractors. Their furniture featured small nails for pulls. Fretwork and inlaid strips were used for decoration. The line of furniture was called H.M. Miniatures. They continued to supply furniture to Block House for nearly twenty years.

Rooms of furniture offered by H.M. Miniatures included a Queen Anne Bedroom, Eighteenth Century Living Room, Duncan Phyfe Dining Room, Chippendale Bedroom, Empire Bedroom, and Modern Bedroom. Most of these designs continued in the line until the end of production in the 1960s. Prices circa 1950 for the modern bedroom pieces ranged from 65 cents for the night table to $1.25 for the twin bed. A sofa was priced at $3.00 and a grandfather clock sold for $2.00.

Furniture pieces sold by Block House circa 1948. These items were part of the line redesigned by Henry Messerschmidt in the mid 1940s. Pictured along with their boxes are the Empire vanity and bench ($50-60), Empire Chifforobe ($50-60), and 18th Century end table ($18-22). *Furniture from the collection of Dian Zillner. Photograph by Suzanne Silverthorn.*

Other Lynnfield kitchen pieces included a table ($20-25), chairs ($10-12 each), and a cabinet ($30-40). Also made for the kitchen were a breakfast nook, and another cabinet. *Furniture from the collection of Mary Lu Trowbridge. Photograph by Bob Trowbridge.*

Lynnfield 1" scaled nursery set circa 1940. Several of the pieces are marked "Made in U.S.A." Furniture included a potty chair ($25-30), chest of drawers ($30-35), table ($20-25), chair ($15-20), and crib ($30-40). *Furniture and photograph from the collection of Lois L. Freeman.*

Other Wood Dollhouses and Furniture 131

Most of this furniture was pictured in the Block catalog circa 1950 as part of the Empire bedroom set. Included were the vanity and stool ($50-60), chifforobe ($50-60), night table ($20-25), mirror ($15-20), and dresser ($50-60). These beds were made for many years with minor variations. The beds pictured ($90-100 pair) differ slightly from those shown in the catalog. *Furniture and photograph from the collection of Judith Armitstead.*

Furniture sold by Block House around 1950. The furniture was advertised as being made of "Genuine Mahogany" and was also scaled 1" to one foot. The rocking chair was sold as part of the Chippendale bedroom ($45-50). The other furniture was pictured in the catalog as part of the 18th Century Living Room. Included were the secretary ($90-120), Duncan Phyfe Sofa ($75-95), drum table ($45-55), and upholstered footstool ($18-22). *Furniture and photograph from the collection of Judith Armitstead.*

Around 1964 the Messerschmidts retired and the production of the popular dollhouse furniture ended. According to Dee Snyder, (*Nutshell News*, May 1980), Richard Roeder was contacted by F.A.O. Schwarz to see if he could manufacture the furniture in South America. Under the name Andi Imports, the new company was begun with a factory in Colombia. Most of the furniture was sold by F.A.O. Schwarz or Block House. The company was only in business two or three years. The mostly mahogany furniture was made in the earlier designs, but in this case, each piece was marked with a sticker which read "Made in Columbia." The Duncan Phyfe Dining Room, Empire Bedroom, and 18th Century Living Room all remained in the line. Extra pieces included a grandfather clock, fireplace, grand piano, drum table, desk, secretary, rocker, and bookshelf. The small nail pulls were replaced by die-stamped metal pulls.

Dee Snyder's article describes how Sonia Messer became involved with the miniature furniture line. According to Snyder, Messer heard about the dollhouse furniture factory during a trip to Brazil in the mid 1960s. Although she had formerly been the operator of a travel agency, she became interested in this new project and contracted to buy all the furniture that the factory could produce. She contacted Block House to be her representative in the East and she became her own distributor in the West. For six years Block House sold her designs in the East under their own name. The furniture included the old Messerschmidt Federal and Duncan Phyfe designs as well as her new lines.

The first of the original Messer designs was a Queen Anne Colonial set which included thirty-seven pieces. With the success of this new furniture, Messer produced many more lines in the years to come. Included were the Queen Anne, 18th Century French, Early American, Italian 17th Century, Victorian Rococo, Spanish, Contemporary Danish, Federal, and Country

This Duncan Phyfe dining room is also circa 1950. The buffet pictured in the Block catalog has regular brass pulls while this one has the earlier nail pulls. Included in the set were a buffet, corner cupboard, extension table, chairs, arm chairs, and a side table (Set $375-$450). *Furniture and photograph from the collection of Judith Armitstead.*

English sets. Each design included enough pieces to furnish a bedroom, living room, and dining room. The Sonia Messer furniture was marked "Made in Columbia/Exclusively for Sonia Messer Imports/Los Angeles, California." The only pieces that were not marked in this manner were the early ones distributed by Block House, Inc. under their own name.

The Sonia Messer furniture retained the high quality first begun by Charles Waite nearly fifty years earlier. As the years passed, the cost of producing this type of miniature, in competition with products made much more cheaply in the Far East, became too difficult and production ended in the early 1980s.

As collectors find it harder to locate the truly old dollhouse furniture, many are turning to the pieces which began as Lynnfield and ended as Sonia Messer. Because of the length of time the furniture was produced, many of the later items are still available through dealers of antique and collectible dollhouse furniture.

132 Other Wood Dollhouses and Furniture

This "Swedish Blonde Maple Finish" modern dining room was featured in the Chestnut Hill catalog in 1951. Each piece was sold individually. Prices ranged from $1.60 for a side chair to $4.50 for the buffet (which also included a separate round mirror). The set included an extension dining room table with inlaid table-top, a shelf server, silver chest, side chest, arm chairs, side chairs, and the buffet with mirror. Pictured are the table ($40-50), shelf server ($30-40), buffet ($60-70), side chest ($40-50), arm chair ($10-15), and side chair ($10-12). A matching modern bedroom was also offered in the same catalog. *Furniture from the collection of Bob Milne. Photograph by George Mundorf.*

Another later version of the Empire bedroom. This furniture dates from the early 1960s. The original Block House prices ranged from 95 cents for the bed tray to $4.50 for the vanity and mirror. Included in the set were the dresser, vanity with mirror, chifforobe, upholstered chair, night table, twin beds with mattresses, mirror, bench for vanity, and bed tray (Set $225 and up). *Furniture and photograph from the collection of Roy Specht.*

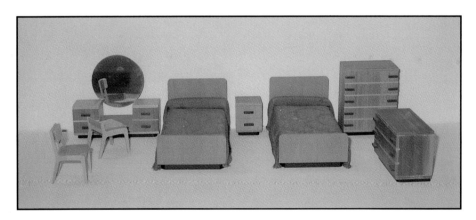

In 1962 Block House marketed a light maple modern bedroom set based on an original Lynnfield design that had been modified somewhat from the original. Included were a dresser with mirror, vanity with mirror and bench, chifforobe, upholstered chair, night table, and twin beds with mattresses. The modern dining room was also featured the same year (Set $225 and up). *Furniture and photograph from the collection of Jeanne Kelley.*

Although some retailers were still labeling the furniture "Lynnfield" during the early 1960s, by the later years of the decade the furniture was being referred to as "Columbian." Pictured is a Columbian dining room set from this period (Set complete with side table $225-255). *Furniture and photograph from the collection of Roy Specht.*

Other Wood Dollhouses and Furniture 133

Additional pieces of furniture marked with stickers, "Made in Columbia" dating from the mid 1960s. Shown are a clock ($15-20), fireplace ($22-25), piano with music box ($85-100), and secretary ($60-75). *Furniture from the collection of Dian Zillner. Photograph by Suzanne Silverthorn.*

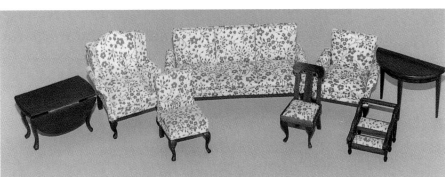

When Sonia Messer took over the designing of the furniture made in Columbia, she continued to produce the earlier designs including the Federal Dining Room and the Queen Anne Living Room. Pictured is the living room from the 1970s which had undergone some changes since the earlier years. Additional items included a coffee table, footstool, desk, table, and tea cart. This set is marked "Made in Columbia/Exclusively for/Sonia Messer Imports/Los Angeles, California." (Chairs $15-20, sofas $30-40, and tables $15-25). *Furniture from the collection of Dian Zillner. Photograph by Suzanne Silverthorn.*

Sonia Messer also designed new lines of furniture to be produced in Columbia. Pictured are the living room pieces from the Queen Anne English Deluxe set of living room furniture. Seventeen pieces were in the complete set which included a sofa, high-back chair, love seat, arm chair, foot stool, coffee table, tea table (2), secretary, desk chair, grandfather clock, tea cart, lowboy, sewing cabinet, swing-leg table, grand piano, and piano bench (Range of $20-25 for a chair or clock to $65-75 for a sofa, piano, desk, or book shelves). *Furniture and photograph from the collection of Roy Specht.*

Sonia Messer Federal bedroom dating from the 1970s. A vanity table, bench, rocker, bedroom chair, and bedroom table were also available to complete the set (Large items $30-35, night stand and mirror $12-15 each). *Furniture from the collection of Gail and Ray Carey. Photograph by Gail Carey.*

134 Other Wood Dollhouses and Furniture

The Queen Anne English DeLuxe dining room line included a table, six dining chairs, host chairs, buffet with fireplace, china cabinet, round occasional table, occasional chair, and tilt-top table (Set $245 and up). *Furniture and photograph from the collection of Roy Specht.*

Another original design by Sonia Messer was the 18th Century French line dating from 1970. Included in the living room were a sofa, upholstered chair, lamp tables, coffee table, console table, console mirror, lady's desk, round-back occasional chair, Bergere chair, grand piano, piano bench, poodle lounge, clock, and round tea table. Missing from the photograph are the piano, bench, and clock (Range from $15-20 for the poodle lounge to $65-75 each for the sofa or desk). *Furniture and photograph from the collection of Roy Specht.*

The Queen Anne English Deluxe bedroom furniture included a bed, night tables, corner chair, dresser with mirror, chifforobe, powder table, vanity bench, chaise lounge, screen, and occasional table (Range from $15-20 for a bedside table to $65-75 for the powder table with mirror). Originally these two items sold for $14.95 and $42.95 respectively. *Furniture from the collection of Gail and Ray Carey. Photograph by Gail Carey.*

MENASHA WOODENWARE CORPORATION

The Menasha Woodenware Corporation, located in Menasha, Wisconsin, has been in business since 1849. It was begun by Elisha D. Smith to produce wooden barrels, tubs, pails, and other woodenware. The company is still operating but its main product today is corrugated cardboard and plastic storage boxes. It is now known as Menasha Corporation.

During the depression the company needed to sell new products in order to stay in business and they began producing toy furniture made of wood. Menasha manufactured several products under the Tyke Toys trade name, including bassinets, tables and chairs, shoo flys, nursery chairs, and dollhouse furniture.

Because the company owned so much woodworking machinery (unlike other smaller manufacturers) they were able to produce a very nice line of 1" to one foot scale dollhouse furniture with turned legs and original panel designs. When the furniture was advertised in *Playthings* in August 1934 the company billed itself as the largest manufacturer of turned woodenware in the world. Furniture was produced for a living room, bedroom, kitchen, dining room, and bathroom. Most of the furniture can be recognized because of its distinctive pattern of turned legs.

The living room pieces included a sofa, footstool, chair, library table, radio (on short legs), table lamp, floor lamp, lamp table, and grand piano. The bedroom included twin beds, vanity, stool, and a round night table. A chest of drawers may also have been made. The bathroom pieces included a bathtub, sink, toilet, stool, and medicine cabinet. The medicine cabinet had a round hole in its top and the door opened. Originally the bathtub included a wood wall with a tile design. The kitchen and dining room chairs were apparently made from the same pattern. The kitchen chairs were painted cream or green while the dining room furniture was finished in either a wood stain or a dull red. Several different tables were produced in the furniture line. These included a drop leaf table, a six legged table, and a blue square table to match the living room furniture. Other dining room items included a buffet with two drawers, a server, and a hutch that may have been meant for the kitchen. The kitchen stove was made on legs with black stencil designs to indicate doors and burners. A sink and probably a refrigerator were also produced for the kitchen. In addition to the 1" scale furniture, a few larger pieces were made during this time period. The rocking chair was 6" tall and was constructed almost entirely of turned wood as was a chair in the same scale. The Menasha furniture is very hard to locate because the company did not stay in the dollhouse furniture business for very long.

Another style of Menasha living room chair had a more modern look ($20-25). *Chair from the collection of the Neenah Historical Society. Photograph by E. Munroe Hjerstedt.*

Menasha Woodenware Corporation advertised their Tyke Toys in *Playthings*, August 1934. Besides dollhouse furniture, the company also made baby bassinets, tables and chairs, nursery chairs, and Shoo Flys. *Photograph by Suzanne Silverthorn.*

Wood dollhouse furniture was made by the Menasha Woodenware Corporation of Menasha, Wisconsin, during the 1930s. The furniture was scaled 1" to one foot. Besides the items shown, the living room pieces included a table lamp ($20-25 each). *Furniture and photograph from the collection of Patty Cooper.*

136 Other Wood Dollhouses and Furniture

The Menasha wood kitchen stove was painted green with black decoration ($20-25). *From the collection of the Neenah Historical Society. Photograph by E. Munroe Hjerstedt.*

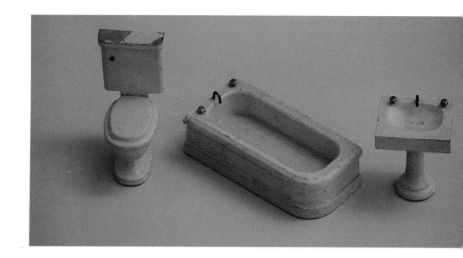

The Menasha Tyke Toys bathroom furniture included a bathtub, toilet, and sink ($20-25 each). *From the collection of Dian Zillner. Photograph by Suzanne Silverthorn.*

A Menasha drop leaf table can be easily identified by the unusual turnings of its legs. The same design was used on the legs of most Menasha furniture ($20-25). *Table from the collection of the Neenah Historical Society. Photograph by E. Munroe Hjerstedt.*

A cabinet and a sink were also part of the pieces made by Menasha Woodenware for the kitchen ($20-25 each). *Childhood toys of Mildred Felder Whitmire. Photograph by Hart Felder.*

The bedroom furniture made by the Menasha Woodenware Corporation was painted green and cream (Large pieces $20-25, small $10-15). *Furniture and photograph from the collection of Patty Cooper.*

The Menasha dining room pieces, like all Menasha items, include functional drawers ($20-25 each). *Furniture and photograph from the collection of Patty Cooper.*

MYSTERY HOUSES AND OTHER HOUSES SOLD BY F.A.O. SCHWARZ

The first F.A.O. Schwarz toy store was opened in 1862 in Baltimore, Maryland. The store was founded by Frederick August Otto Schwarz and his three brothers. By the turn of the century, the headquarters were moved to New York and the firm has been a fixture on Fifth Avenue for many decades. The company now operates 38 stores across the United States. During their years in business, F.A.O. Schwarz has made an effort to carry toys not sold by any other company. Sometimes this policy meant only that a popular toy firm offered one of their usual dolls with a different suitcase full of clothing specifically produced for Schwarz customers. Other original toys were made exclusively for the F.A.O. Schwarz market. This was certainly true of many of the dollhouses and furnishings carried by their toy stores over the years.

In the late 1800s, Schwarz carried a line of wood dollhouses known to collectors as "Mystery Houses." There were several different designs of these houses but most can be easily identified because of the unusual strips of wood used on the outside for decoration. Flora Gill Jacobs, in her book *Dolls' Houses in America*, has verified through an advertisement that these houses were carried by F.A.O. Schwarz in 1897. The house pictured in the ad was 45" high x 35" wide x 24" deep. It was priced at a very expensive $33. If the firm did contract to have the houses made exclusively for them, there appears to be no record of who made the houses or for what length of time they were produced. The two-story houses were quite large and were made with four to six basic rooms. In addition, some contained halls and attic rooms as well. The larger houses featured dormers and an extra wing and the base on these models could be over five feet wide. The houses were quite well made and featured parquet floors, detailed woodwork, and paneled doors.

During the 1930s, Schwarz sold several different models of Tudor dollhouses. Like the Mystery Houses, there was a variety of these houses carried in the Schwarz catalogs throughout the decade. In 1932 a cottage was featured with only three rooms but by 1937 the house had been enlarged to a six room model. Both houses featured the same stucco finish, the same style metal windows, and a trellis on the front wall. Plans for the larger house were also advertised in the *Popular Homecraft* magazine in the November-December 1938 issue with information that the house had been sold by a "large department store" the preceding year. The store price was listed at $65 for the house equipped with electric lights. It is not known if one person or several people supplied Schwarz with houses made from specific plans through the years. Flora Gill Jacobs, in her book *Dolls' Houses in America* identified a similar house with an attached "Toy Crofters" label as having been purchased from Schwarz. Toy Crofters was located in Beacon, New York, and it may have been a small firm that produced custom items for F.A.O. Schwarz.

Other Schwarz houses now valued by collectors include an exclusive wood model sold by the toy company from the late 1950s until the early 1970s. The large house was advertised in the Schwarz catalogs and was also featured in their many stores. The house contained seven rooms and was 35" wide x 26" high x 20" deep. The house was scaled 1" to one foot. The front of the house was removable and the back was open for easy access. The house was two rooms deep except for the living room and one bedroom which extended from the front to the back. The house was constructed of white pine plywood and included a distinctive two story porch on one end and electric lights. F.A.O. Schwarz carried the Colombian (see Lynnfield-Block House) furniture which could be purchased to furnish the house.

Collectors call this house one of the "Mystery Houses" that was sold by F.A.O. Schwarz in the late 1890s. The houses included unusual strips of wood decoration on the outsides and unique woodwork around the windows. This house was priced at $33 by Schwarz in 1897 ($6,000-10,000). 40" tall x 33" wide x 20" deep. *House from the collection of Leslie and Joanne Payne. Photograph by Leslie Payne.*

The inside of the smaller Mystery Houses included four rooms in two stories. Some of these houses had one dormer window while others had two. The house has been repapered. 40" tall x 33" wide x 20" deep. *House from the collection of Gail and Ray Carey. Photograph by Gail Carey.*

A larger Mystery House was made with one wing. These two-story houses contained six rooms. The houses included paneled doors and parquet floors (Not enough examples to determine a value). 40.5" high x 42.5" wide x 18" deep. *Courtesy of the Toy and Miniature Museum of Kansas City.*

The largest of the Mystery Houses was over six feet wide and contained eleven rooms including halls and attic areas. According to the owner, this house was sold in 1906 by F.A.O. Schwarz in New York City (Not enough examples to place a value). *House and photograph from the collection of Anne B. Timpson.*

The Timpson house has been furnished in the style of the 1890s and contains bamboo furniture and gilt metal pieces as well as furniture made by the German Marklin firm. *House and photograph from the collection of Anne B. Timpson.*

Although this house does not have the unusual wood trim of other "Mystery Houses," it is attributed to the same maker. With its clapboard siding, its style is more like a colonial mansion (Not enough examples to determine a value). 59" high x 60" wide x 37" deep. *Courtesy of the Toy and Miniature Museum of Kansas City.*

Other Wood Dollhouses and Furniture 139

The inside of the colonial house contains six rooms plus two halls and attic space. This house also includes the Mystery House woodwork around the inside windows and paneled doors. *Courtesy of the Toy and Miniature Museum of Kansas City.*

A Tudor style house was sold by the F.A.O. Swartz company during the late 1930s. The house has a "stucco" finish, metal windows, and a trellis on the front wall ($1200-1500). 29" high x 42" wide by 17.75" deep. *House and photograph from the collection of Lois L. Freeman.*

The large Tudor house contains six rooms plus an upstairs hall. A pattern for this house was featured in the *Popular Homecraft* magazine from 1931 in the November-December issue. The house featured electric lights. *House and photograph from the collection of Lois L. Freeman.*

This model Schwarz house was sold by the company from the late 1950s until the early 1970s. It contains seven rooms and is scaled 1" to one foot. The side porch makes the house easy to identify. The chimney and steps are missing ($900-1,000). 26" high x 35" wide x 20" deep. *House from the collection of Dian Zillner. Photograph by Suzanne Silverthorn.*

The house is two rooms deep and when the front is removed, the living room, dining room, two bedrooms, and the bathroom can be accessed. The kitchen, nursery, hall, living room, and large bedroom can be reached through the open back. This house, like many of the originals, is furnished with Columbian and Sonia Messer Furniture. It has been redecorated. *House from the collection of Dian Zillner. Photograph by Suzanne Silverthorn.*

WISCONSIN TOY COMPANY

The Wisconsin Toy Company was located in Milwaukee, Wisconsin, during the 1920s and 1930s. The firm produced both wood dollhouse furniture and dollhouses during this period. The company was listed in the Milwaukee city directory from 1921 to 1936. The company president was Frank R. Wilke in 1921 but Ernst H. Johannes was listed as president from 1922 until 1935. The company was housed in several different locations in the city during its residence in Milwaukee.

The furniture made by Wisconsin Toy ranged from a very large 1" to one foot scale to a regular 1" to one foot scale. The trade name used for the furniture was "Goldilocks." Some of the pieces are marked in this manner and other items are marked in a triangle "Wis/Toy/Co." Most of the furniture carries no company marking although some pieces have the item number penciled on the bottom giving the collector an additional clue.

A catalog from the mid 1930s pictured many different rooms of furniture in a variety of styles. Four different bedrooms were shown ranging from a new modern "Art Deco" design to one with a more traditional look. Living rooms were pictured with either upholstered pieces or wood sofas and chairs. A variety of kitchen items were offered including an ironing board that folded out from a cabinet as well as a sink made of plaster. No kitchen stove was shown in the set. One of the nicest Wisconsin rooms of furniture was the nursery which included a high chair, nursery chair, bassinet, dressing table, and nursery dresser. This set sold for $2.50. Breakfast sets and dining room pieces were also included in the catalog. The bathroom fixtures, like the kitchen sink, were made of a plaster or chalk-like material. Many different finishes were used on the various rooms of furniture including enamel, stain, and varnish.

Three different houses were pictured in the catalog. All were open on the front and contained six to eight rooms. Some of the houses also included electric lights. The smallest house was 32" tall x 32" wide x 18" deep. Earlier furniture thought to be produced by Wisconsin Toy was a little larger in scale and many pieces were more crudely made. It could be that the company started as a very small firm with the original furniture being produced by workers not yet familiar with the manufacture of dollhouse furniture. As the quality improved, the designs became more complicated. Eventually the furniture was carried by large department stores like Stix, Baer, and Fuller Co.

Because the furniture came in so many styles including Louis XIV, Early American, Art Moderne, Modern, and Traditional, the pieces appeal to many collectors. It may be difficult to find enough furniture to furnish an entire house with "Wisconsin" but the current collector may be lucky enough to locate pieces for a room or two and much of the later furniture would fit nicely with Tynie Toy items.

This bedroom furniture came with a chair that is shown in the Wisconsin Toy Catalog from the mid 1930s and may be an earlier set of the company's bedroom furniture. The set was also made in white. The furniture is in a large 1" to one foot scale and the drawers are functional (Chair $15-20 other pieces $25-35 each). *Furniture from the collection of Dian Zillner. Photograph by Suzanne Silverthorn.*

Other Wood Dollhouses and Furniture 141

The early bedroom pieces were made to accompany either a large bed or a smaller bed ($25-35 each). *Furniture and photograph from the collection of Patty Cooper.*

This set of kitchen furniture includes a Wisconsin Toy table shown in a 1930s Wisconsin catalog. The other pieces may date from an earlier kitchen set sold by the company. All are in a large 1" scale. The sink is made of wood instead of plaster ($30-40). Other items include an ice box ($20-25), chair ($10-15), and table ($25-30). *Furniture and photograph from the collection of Patty Cooper.*

Overstuffed living room furniture in the 1" to one foot scale stamped "Goldilock's" circa mid 1930s. This trade name was used by the Wisconsin Toy Co. located in Milwaukee, Wisconsin. Shown are a sofa ($45-55), chair ($30-35), and ottoman ($10-15). *Furniture and photograph from the collection of Lois L. Freeman.*

Various pieces of Goldilock's wood stained 1" scale furniture from the 1930s. Included are a telephone table and bench ($25-30), two chairs ($25-30 each), and a davenport table ($25-30). *Furniture from the collection of Dian Zillner. Photograph by Suzanne Silverthorn.*

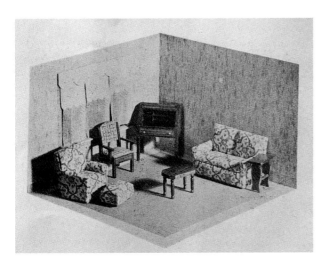

Also pictured in the company catalog in the "Lounging Suite" was a pull up chair, end table, writing desk, and coffee table. The complete set sold for $4.00 while the individual pieces were priced from 28 cents to $1.50 each. *Catalog from the collection of Leslie and Joanne Payne. Photograph by Suzanne Silverthorn.*

142 Other Wood Dollhouses and Furniture

Wisconsin Toy Co. upright and grand pianos. Each is marked with the "WisToyCo" triangle stamp. The upright piano also includes a paper price label from Stix, Baer, and Fuller Co. for $1.25 ($45-65 each). *Furniture and photograph from the collection of Lois L. Freeman.*

Kitchen furniture marked Wisconsin in 1" scale. Included are a wall hung plaster sink ($35-45), a two door wood ice box ($25-30), kitchen cabinet ($65-75), table ($25-30), chairs ($20-25 each). A high stool was also pictured in the company catalog to accompany this set. *Furniture from the collection of Ray and Gail Carey. Photograph by Gail Carey.*

Wisconsin Toy Company also pictured another kitchen set in their 1930s catalog. Included were this table ($25-30), and two of these kitchen cabinets ($35-45), as well as a chair, sink (with bottom cupboard), stool, and folding ironing board. *Furniture from the collection of Ray and Gail Carey. Photograph by Gail Carey.*

This 1" scale cabinet with ironing board and stool ($30-35) were part of the more modern kitchen set pictured in the Wisconsin Toy catalog from the 1930s. *Furniture and photograph from the collection of Patty Cooper.*

This Wisconsin breakfast nook called "Goldilock's Pullman Diner" sold for $1.00 during the 1930s. *Catalog from the collection of Leslie and Joanne Payne. Photograph by Suzanne Silverthorn.*

Other Wood Dollhouses and Furniture 143

This Goldilock's bedroom suite was also pictured in the Wisconsin catalog. This chest of drawers is exactly like the one in the earlier bedroom set except the top has been made smaller to give the piece a more modern look and a decal has been added ($10-15 each for small items, $25-35 each for larger pieces). *Furniture and photograph from the collection of Patty Cooper.*

Other items in the original set include a two drawer dresser and a rocking chair. A smaller bed and a folding screen could be purchased separately. This repeats the concept of two sizes of beds offered in the earlier set of bedroom furniture. The dresser is also like the earlier one with the addition of a flush top and a newly designed plain mirror frame. *Catalog from the collection of Leslie and Joanne Payne. Photograph by Suzanne Silverthorn.*

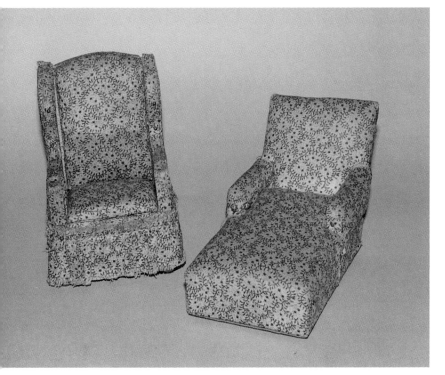

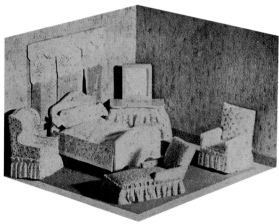

This Wisconsin "Early American Chintz Bedroom Furniture" also dates from the 1930s. The bedroom pieces included an armchair, vanity with bench, chest of drawers, bed, mattress, wing chair, and chaise lounge. *Catalog from the collection of Leslie and Joanne Payne. Photograph by Suzanne Silverthorn.*

Chaise lounge and chair from the Wisconsin "Chintz Bedroom." Both pieces are marked "Goldilocks". The chaise is missing the bottom ruffle ($25-35 each). *Furniture and photograph from the collection of Lois L. Freeman.*

This unusual "Moderne Bed Room Set" was also pictured in the 1930s Wisconsin Toy catalog. The five piece set included a dresser, vanity, bench, chair, and bed. *Catalog from the collection of Leslie and Joanne Payne. Photograph by Suzanne Silverthorn.*

144 Other Wood Dollhouses and Furniture

Wisconsin Toy dining room suite. The complete set of furniture originally sold for $4.00 in the company catalog and included three straight chairs and a host chair ($15-20 each), table with leaves ($30-40), server ($25-30), buffet ($35-40), and china cabinet ($40-50). *Furniture from the collection of Zelma Fink. Photograph by Suzanne Silverthorn.*

One of the most desirable sets of Wisconsin furniture was the nursery. Included were a high chair, potty chair, crib, dressing table, and nursery dresser ($35-45 each). *Furniture and photograph from the collection of Patty Cooper.*

In addition to the regular dining room table, Wisconsin Toy also produced a gate leg table ($30-35), that they offered as part of the "Occasional Set." Other items included a console table, console mirror, secretaire, pull up chair, tilt top table, and highboy. The furniture came finished in mahogany or maple. *Furniture and photograph from the collection of Patty Cooper.*

Besides furniture, Wisconsin Toy also produced several different models of dollhouses. These houses varied in size and contained from six to eight rooms. The larger houses had stairways, and some of the dollhouses featured electric lights. The larger houses complete with furniture were priced at $62.50, which was quite expensive during the depression years. *Catalog from the collection of Leslie and Joanne Payne. Photograph by Suzanne Silverthorn.*

The Wisconsin Toy bathroom set is very unusual since it is made of a plaster-like material. These pieces are all stamped "Goldilock's" (Fixtures, $40-50 each, medicine cabinet $20-25). *Furniture and photograph from the collection of Lois L. Freeman.*

MISCELLANEOUS WOOD DOLLHOUSES (AND SOME FURNITURE)

Arrow Handicrafts Corp.
See Little Orphan Annie Dollhouse

Cass Co., (N.D.)

The N.D. Cass Co., located in Athol, Massachusetts, celebrated its one hundredth anniversary in 1996. The firm was founded in 1896 by Nathan David Cass and it has been a family owned company throughout its history. The present CEO is William Cass Jr., a grandson of the founder. The current president of N.D. Cass is the great grandson of the original N.D. Cass.

The Cass firm has been involved with making toys for many years, but its first product was suitcases. Folding beach chairs have also been a profitable product for the company through the years. In the toy line Cass has produced doll trunks, doll furniture, and dollhouses as well as other products.

An advertisement in *Playthings* from 1910 pictured oak doll furniture including a table, four chairs, a rocking chair, settee, and another style table and chair. Although the ad did not include the size of the furniture, the same table and chairs were pictured in a Montgomery Ward catalog in 1916 with chairs nearly 7" tall. It may be that some of the company's furniture was also made in a smaller scale that could be used with a large dollhouse.

The Cass company did not keep early catalogs and the oldest one now on file dates from 1928. This catalog contains a picture of a one-room wood bungalow. The house had the architectural details printed directly on the wood instead of on applied lithographed paper. The same design was used on the front of a house shown in a 1926 Morton E. Converse catalog, but the houses had different roof styles, chimneys, and foundations. It may be that both companies bought the designs from another contractor and then assembled the dollhouses at their own plants. The Cass bungalows were produced in six different sizes from 10 1/4" x 9" x 8" to 20 1/2" x 18" x 15". The houses opened from the side. The similar Converse houses came in five sizes from 8 3/4" x 8 3/4" x 8 1/2" to 17" x 17" x 15 1/2". The Cass company also advertised "stucco" dollhouses circa 1930 but the company has no record of these houses.

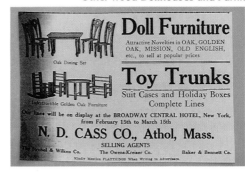

The N.D. Cass Co. advertisement from *Playthings* magazine in 1910 featured doll furniture made by the company. Although measurements are not listed in the ad, the table and chairs were pictured in a Montgomery Ward catalog in 1916 with chairs nearly 7" tall. Wards sold the set for 21 cents. *Photograph by Suzanne Silverthorn.*

Oak dining room furniture circa 1914 which is approximately 1 1/2" to one foot in scale. Although it is not known that the furniture was made by Cass, the chairs have the same curved back as pictured in the ad. The furniture is very similar to that made by the Star Novelty Works but the Star chairs do not have curved backs (Set $200 and up). *From the collection of Dian Zillner. Photograph by Suzanne Silverthorn.*

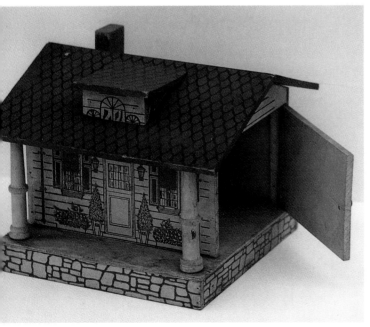

Cass featured this wood house in their 1928 catalog. The house has the lithographed design printed directly on the wood and is similar to houses produced by the Morton E. Converse firm ($300-350). 9" high x 10.25" wide x 8" deep. *From the collection of Dian Zillner. Photograph by Suzanne Silverthorn.*

The chairs on this oak set of furniture have the same legs as the chairs on the dining room furniture. However, the backs of these chairs are straight instead of curved. The furniture is approximately 1 1/4" to one foot in scale (Set $30-40). *From the collection of Dian Zillner. Photograph by Suzanne Silverthorn.*

Other Wood Dollhouses and Furniture

Converse, (Morton E.)

The Morton E. Converse and Son firm had its beginning in a company called Mason and Converse which was founded in Winchendon, Massachusetts. In 1884 Converse formed his own company under the name of Morton E. Converse Co. In 1898 the name was changed to Morton E. Converse and Son. Converse was a leader in the manufacture of wood dollhouses in the United States. Most of the houses were of the bungalow type with a stone foundation printed on the wood. The other architectural details were also lithographed directly on the wood. The houses are often confused with those produced by the N.D. Cass Co. Some of the Converse houses were finished in blue and red while others were done in red and green. In 1931 the company produced a very nice line of wood 3/4" to one foot scale dollhouse furniture. At the same time, the company marketed a Realy Truly four-room two-story dollhouse made of cardboard and wood. See *American Dollhouses and Furniture From the 20th Century* for more information on the Converse dollhouses and furniture.

The Converse house is front-opening and contains two decorated rooms. The Converse name can be seen at the edge of the downstairs "rug." *House and photograph from the collection of Patty Cooper.*

This wood two-story dollhouse, made by Morton E. Converse, also has the lithographed design printed directly on the wood ($700-800). 18" to roof top x 10.25" wide x 7" deep. *House and photograph from the collection of Patty Cooper.*

Other Wood Dollhouses and Furniture 147

Although this wood house is not marked, it has been attributed to Converse. The architectural design is printed directly on the wood. The side porch adds an unusual touch ($450-550). 11.25" high x 11.75" wide x 7" deep. *House and photograph from the collection of Rita Goranson.*

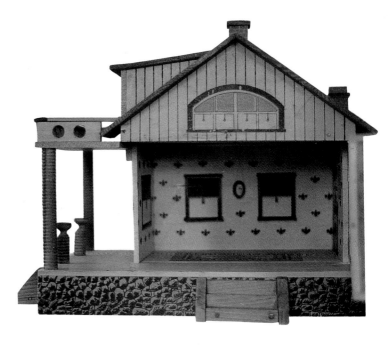

The inside of the house contains only one room but the decoration is nicely printed on the wood. *House and photograph from the collection of Rita Goranson.*

In 1931 Converse advertised their Realy Truly dollhouse which appears to be nearly identical to this four-room house. It was made of fiberboard reinforced with wood according to the advertisements. The ad did not mention that the house had a wood roof as this house does. The house pictured in the company advertisement also included a front section of dormer windows. The house was marketed to accompany its Realy Truly furniture then being manufactured ($400-450). 15.75" high x 19" wide x 10" deep. *House from the collection of Becky Norris. Photograph by Don Norris.*

The Realy Truly wood 3/4" to one foot scale bedroom furniture was produced by Converse circa 1931. The company also made pieces to furnish a living room, kitchen, and dining room ($25 each piece). *Furniture from the collection of Dian Zillner. Photograph by Suzanne Silverthorn.*

Keystone

The Keystone Manufacturing Company was founded in Boston in the early 1920s by Chester Rimmer and Arthur Jackson. At that time the company produced movie projectors. From about 1940 until 1950 the firm manufactured dollhouses. The houses were made from Masonite with most of the houses decorated both inside and out. Many of the large houses featured a curved staircase and a closet. Some of the houses are marked while others are not. See *American Dollhouses and Furniture From the 20th Century* for more information on Keystone houses.

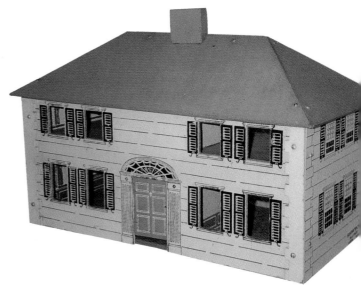

Several different designs and sizes of this Keystone house were produced. The roofs were not all made in the hipped style and the inside decorations varied somewhat but the basic green and blue coloring remained constant. Unlike most of these houses, this one is marked "Keystone Mfg. Co./288 A Street/ Boston, Mass." ($65-75). 13" high x 19.25" wide x 9" deep. *House and photograph from the collection of Marilyn Pittman.*

The inside of the Keystone house was decorated in green and blue but the floors were plain. The windows are usually missing from these houses. *House and photograph from the collection of Marilyn Pittman.*

This Tudor house circa 1941 features metal windows and shutters found on several marked Keystone houses. The front chimney is unusual in Keystone houses ($175-200). 16.5" high x 25" wide x 10.5" deep. *House and photograph from the collection of Marilyn Pittman.*

Other Wood Dollhouses and Furniture 149

This house with a similar chimney and front door treatment is also attributed to Keystone ($175-200). 16" high x 24" wide x 9" deep. *House and photograph from the collection of Lois L. Freeman.*

The interior of the unmarked Masonite four room house attributed to Keystone is quite plain with no interior decoration. *House and photograph from the collection of Lois L. Freeman.*

This Keystone Masonite house contains four rooms and features metal windows, wood planters, wood porchlight, and one remaining awning which gives the house a different look. The inside of the house is quite plain but includes electric lights and a built-in fireplace. House circa 1940s. The house is marked with the Keystone name on its side.($125-135). 20" to the top of the chimney x 25" wide x 9" deep. *House from the collection of Dian Zillner. Photograph by Suzanne Silverthorn.*

Kiddie Brush and Toy Co.

The Kiddie Brush and Toy Co. was founded in 1930 by Paul A. Jones Sr. and John Doty in Bryan, Ohio. The company moved to Jonesville, Michigan, in 1932 when Jones bought out his partner's interest. The trade name "Susy Goose" came into use for the firm's toys soon after the move to Michigan.

The earliest toys made by the company were housekeeping pieces which included a metal carpet sweeper, floor mops, brooms, and duster.

With the arrival of World War II, the firm devoted its resources to the manufacture of 90MM gun parts. Several B24 Bomber parts were also produced on a contract basis. Although the Susy Goose trade name is probably best known as the brand name for several Barbie™ (Mattel, Inc.) toys, the firm also manufactured dollhouses, a grocery store, and a motel during the 1950s.

Paul A. Jones, Sr. retired from the company in the mid 1960s and the firm continued to operate with the leadership of his son, Paul Jones Jr. until it was liquidated around 1970.

This dollhouse was made by the Kiddie Brush and Toy Co (Susy Goose) in Jonesville, Michigan, circa 1950s. It contains two rooms downstairs and three rooms in the upper story. The partitions upstairs are movable. There is no inside decoration in the house which is made of pressed hardboard. The windows are plastic ($65-75). 19" high x 30" wide x 12" deep. *House from the collection of Dian Zillner. Photograph by Suzanne Silverthorn.*

Little Orphan Annie Dollhouse

Harold Gray was the originator of the "Little Orphan Annie" comic strip which made its debut on August 5, 1924. He worked for the Chicago Tribune-New York News Syndicate. The ever popular strip spawned radio shows in the 1930s and 1940s, a hit Broadway show "Annie" in 1977 and a motion picture, based on that show, made by Columbia Pictures in 1982. It was the merchandising of the Broadway show and movie that produced so many Little Orphan Annie tie-ins. Over sixty-five companies signed up to produce "Annie" merchandise and over 400 different products were made. Of interest to dollhouse collectors is the "Annie" dollhouse from 1978. It came with twenty-five pieces of furniture to furnish the six rooms. Four figures representing Annie, Daddy Warbucks, Sandy, and Punjab were also included. The house was sold in kit form and was to be assembled by the consumer. It was made by Arrow Handicrafts Corp. in Chicago and copyrighted by the Chicago Tribune-New York News Syndicate Co., Inc. The house was called "Annie Dream Dollhouse."

Arrow Handicrafts Corporation located in Chicago marketed the Annie Wood Dream Dollhouse in 1978. The house was to be assembled by the consumer. It came with twenty-five pieces of furniture and four comic figures (Original box $60 and up). 17.25" high x 23.12" wide x 12.75" deep. *House and photograph from the collection of Ruth Petros.*

The inside of the house contained six rooms. *House from the collection of Dian Zillner. Photograph by Suzanne Silverthorn.*

Lundby

Lundby of Sweden, located in Lerum, Sweden, was very successful in marketing their 3/4" to one foot scale furniture and dollhouses during the 1970s and 1980s. The furniture was so popular that it was carried in national mail order catalogs in the United States as well as in the Lundby catalogs distributed throughout the world.

The firm began producing furniture shortly after the end of World War II. The line grew until it consisted of over 200 items by the late 1980s. Furniture was provided for the kitchen, dining room, lounge, bedroom, children's room, bathroom, living room, and laundry. Besides basic items, many accessories and "extras" were offered through the Lundby catalogs. These included lamps, lights, dollhouse dolls, curtains, rugs, pictures, flowers, organs, stereos, fireplaces, and much more.

The Lundby dollhouses were unusual in that they could be purchased in pieces to be combined to complete a house two, three, or four stories tall. The basic hardwood house sold for $79.99 in the Sears Christmas Catalog in 1981. The windows, bannister, and staircase were plastic. Several sets of furniture were also pictured in the catalog with prices ranging from $13.99 for the laundry room to $23.99 for the kitchen pieces. The furniture was machine-tooled hardwood or plastic and featured fabric upholstery.

Lundby purchased the famous English firm of A. Barton & Co. (Toys) Ltd. in 1984. Their products continued to be made and marketed under the Caroline's Home Ltd. trade name. Lundby also continued to make their own products until the firm went out of business during the 1990s.

The 1981 Sears Christmas catalog pictured dollhouses and furnishings made by Lundby of Sweden. The hardwood houses could be purchased in several sections. The furniture was made of hardwood and/or plastic and was scaled 3/4" to one foot. It was sold mostly in room settings priced from $13.99 to $23.99. The largest unfurnished house was priced at $79.99. *Photograph by Suzanne Silverthorn.*

Rich

The Rich Company was founded in 1921 in Sterling, Illinois. The firm originally made tops for automobiles but in 1935 the company moved to Clinton, Iowa, and became the Rich Toy Manufacturing Co. During the 1950s the firm moved to Tupelo, Mississippi. Dollhouses were produced from approximately 1935 to the early 1960s.

Most of the houses were made of U.S. Gypsum hardboard. The earlier houses had acetate windows silk screened with black or white window panes. See *American Dollhouses and Furniture From the 20th Century* for more information on Rich houses.

Rich Cottage circa 1930s with a label which reads "Rich Mfg Co./made from Weatherwood Hardboard/Won't Warp/Clinton, Iowa." The house includes three rooms as well as a garage. The windows are made of a heavy plastic or celluloid material ($100-125). *House and photograph from the collection of Rita Goranson.*

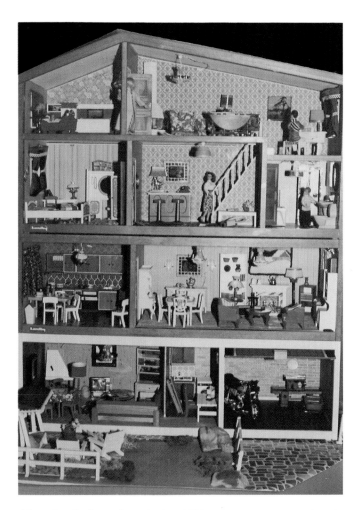

A large Lundby house from the late 1980s contains many examples of the Lundby furniture (House $150-200, furniture $10-12 each). 31.25" high x 26.25" wide x 9.75" deep. *House and photograph from the collection of Mary Harris.*

Other Wood Dollhouses and Furniture

This unusual Rich house dates from the 1930s. The house has the steel corners used on the early houses and includes paper shutters on the outside ($200-225). 18" high x 30" wide x 14" deep. *House and photograph from the collection of Lois L. Freeman.*

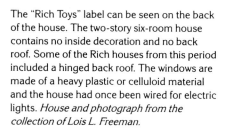

The "Rich Toys" label can be seen on the back of the house. The two-story six-room house contains no inside decoration and no back roof. Some of the Rich houses from this period included a hinged back roof. The windows are made of a heavy plastic or celluloid material and the house had once been wired for electric lights. *House and photograph from the collection of Lois L. Freeman.*

This Rich house also features the steel corners on the front of the house but it is a newer model. The small second floor windows in the front are functional ($200-225). 21" high x 32.5" wide x 14" deep. *House and photograph from the collection of Lois L. Freeman.*

Other Wood Dollhouses and Furniture 153

The inside of the Rich house contains six rooms and a stairway. This house was also wired for electric lights by using the two light strips shown on the back of the house. *House and photograph from the collection of Lois L. Freeman.*

Rich Toys produced several Art Deco houses during their years in business. Although this house is not marked, it is similar to other Deco houses known to have been made by Rich ($350-400). 12" tall x 16.15" wide x 8.75" deep. *House and photograph from the collection of George Mundorf.*

The inside of the Art Deco house contains four rooms in two stories and has no inside decoration. *House and photograph from the collection of George Mundorf.*

Schoenhut

Albert Schoenhut founded the A. Schoenhut Co. in Philadelphia in 1872. The firm was manufacturing dollhouses by 1917. The houses were constructed of fiberboard and wood. Dollhouse furniture was added to the line in 1928. Many different designs were used for both the houses and the two scales of wood furniture. The larger 1" to one foot scale furniture was featured in the 1932 Sears fall mail order catalog. The furniture may have been made exclusively for Sears as it does not appear in the regular Schoenhut catalogs. The design of the furniture is much like that produced by the Strombecker company in 1931. Collectors need to look carefully to distinguish the Schoenhut furniture from the Strombecker pieces. The following pointers should help to make the distinction.

SCHOENHUT

Bedroom
    No holes in headboard.
    Two opening drawers in dresser.
Dining Room
    Green chairs with cut-out backs, but no holes in top.
    Dining table has no stretcher.
Living Room
    Sofa and chair arms affixed by two nails horizontally.
    Ottoman has three layers of cushions. All have turned feet.
Bathroom
    Stall shower.
Kitchen
    Stove has diamond printed on front below knobs.
    Sink has faucet made of bent flat pieces of metal.

STROMBECKER

Bedroom
    Holes in headboard.
    One opening drawer in dresser.
Dining Room
    Chairs have holes at top.
    Table has stretcher.
Living Room
    Sofa and chair arms affixed by two nails diagonally.
    Ottoman has two layers of cushions. All have bead feet.
Bathroom
    Shower sometimes attached to tub.
Kitchen
    Stove has impressed burners.
    Sink has no faucet.

The company went into bankruptcy in 1934 which ended the production of their dollhouses and furniture. Another Schoenhut company was soon established called Schoenhut Manufacturing Co. This firm was headed by Albert F. Schoenhut and his son Fredrick. They sold dollhouses from 1936 to 1940. See *American Dollhouses and Furniture From the 20th Century* for more information on Schoenhut dollhouses and furniture.

The box for this Schoenhut dollhouse reads "Schoenhut/Knock-Down Doll House/Manufactured by the A. Schoenhut Co./Phila. Pa." The four-room house is made of fiberboard and wood and can be taken apart and stored flat ($700-800). 18" high x 20" wide x 13" deep. *House from the collection of Dian Zillner. Photograph by Suzanne Silverthorn.*

Schoenhut wood furniture was first produced by the firm in 1928. The furniture scale was a mixture of large 3/4" to one foot and small 1" to the foot. The living room pieces included several lamps ($25 and up) as well as a fireplace and clock ($50) to supplement the sofa ($30-40), table ($25-30), and chairs ($25-35). *Furniture and photograph from the collection of Patty Cooper.*

Other Wood Dollhouses and Furniture 155

The 1" scaled Schoenhut bedroom furniture included two beds ($40-45 each), two chairs ($20-25 each), dresser with mirror ($40-45), nightstand ($20-25), and floor lamp ($25-30). Unlike the Strombecker pieces, the Schoenhut beds did not have decorative holes in the headboards nor were they marbled. *Furniture and photograph from the collection of Patty Cooper.*

In 1932 the Schoenhut company again produced a larger size furniture when they sold 1" to one foot pieces through the Sears catalog. The box that came with this new furniture pictured the 3/4" to one foot furniture. The dining room pieces included a table, server, buffet, and four chairs. The drawers and doors were functional on this set (Boxed set $300). Unlike Strombecker, the Schoenhut chairs did not have a decorative round hole on the back. *Furniture and photograph from the collection of Patty Cooper.*

The circa 1932 Schoenhut 1" scale living room pieces included a piano and bench ($75-80), chair and footstool ($40-45), side table ($35), library table ($40-45), and sofa ($45-50). This furniture is often confused with the Strombecker pieces of the same period. The Schoenhut library table has no stretcher. *Furniture and photograph from the collection of Patty Cooper.*

156 Other Wood Dollhouses and Furniture

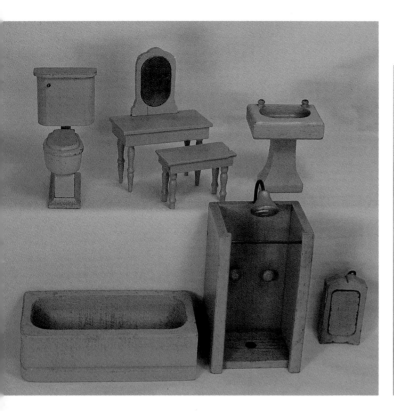

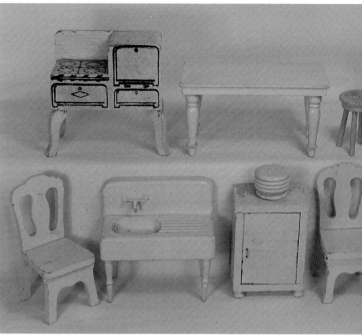

The 1932 Schoenhut 1" scaled bathroom pieces included a shower as well as the more usual furnishings ($45-50). The Strombecker shower from this period was hooked to the bathtub (Fixtures $35-40 each, vanity and stool $50, and medicine cabinet $20-25). *Furniture and photograph from the collection of Patty Cooper.*

The 1" scale Schoenhut kitchen furniture was also similar to the Strombecker pieces but the chairs for the Strombecker kitchen had plain backs. The diamond pattern on the Schoenhut broiler door helps identify the Schoenhut stove (Appliances $45-50 each, table $40-45, chairs $20-25, and stool $20-25). *Furniture and photograph from the collection of Patty Cooper.*

Strombecker

The Strombeck-Becker Manufacturing Co. was established in 1881 in Moline, Illinois. This firm probably produced more wood dollhouse furniture than any company in the United States during its three decades of furniture production (1931-1961). They made many different designs of furniture in order to reflect the changing times. The radio on legs was changed to a floor model which, in turn, became a television in various sets of furniture. The company produced lines of furniture in both the 3/4" to one foot scale as well as the 1" to one foot scale.

Several different houses were also marketed by the company to accompany the furniture. See *American Dollhouses and Furniture From the 20th Century* for much more information on the Strombecker furniture and houses.

The box front of the early Strombecker 1" to one foot scaled wood furniture pictures bedroom, dining room, kitchen, and living room furniture circa 1931. The bathroom pieces were probably added a year later. *Box and photograph from the collection of Patty Cooper.*

Other Wood Dollhouses and Furniture 157

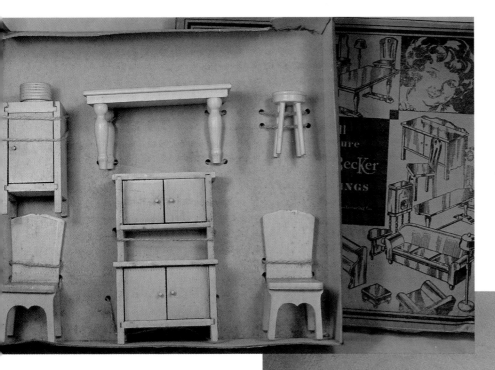

The early 1" circa 1931 kitchen set made by Strombeck-Becker Manufacturing Co. included a table, cabinet, ice box, two chairs, and a stool (Boxed set $150-175). A sink and stove were added later. *Furniture and photograph from the collection of Patty Cooper.*

Strombecker wood dining room furniture circa early 1960s (boxed set $85-100). This set of furniture also came in a bubble pack marked "Strombeck Manufacturing Co." This furniture was made for the open-backed fiberboard dollhouse pictured on the side of the box. *Furniture and photograph from the collection of George Mundorf.*

Susy Goose
See Kiddie Brush and Toy Co.

Tynietoy
The Tynietoy furniture first came on the market when Marion I. Perkins and Amey Vernon opened the Toy Furniture Shop in Providence, Rhode Island, around 1920. The furniture was made in a large 1" to one foot scale and the designs were intended to be miniature reproductions of real antiques. The furniture was crafted with functional doors and drawers. The Tynietoy furniture was available until sometime in the 1950s. After the death of both Perkins and Vernon, the furniture was carried by Louise Fales Specialities for several more years. Most of the furniture was marked with the trademark of a two-story house with a pine tree on the left and a ladder back chair on the right. This trademark was stamped into the wood on most of the furniture.

To accompany the furniture, Tynietoy also carried several different styles of dollhouse dolls. The most popular dolls in their line appeared to be the Peggity dolls. These new jointed dolls were reproductions of the earlier wooden dolls which were made in the 1800s.

Tynietoy offered over one hundred different accessories in their catalogs. Many of the accessories which were carried to complement the furniture came from other sources and were not made by Tynietoy.

The wood Tynietoy Village House contains four rooms ($3,000-4,000). 25" high x 28" wide x 14" deep. *House from the collection of Leslie and Joanne Payne. Photograph by Leslie Payne.*

158 Other Wood Dollhouses and Furniture

The inside of the Tynietoy Village house is filled with Tynietoy wood furniture. A stairway is on the far wall of the living room. *House from the collection of Leslie and Joanne Payne. Photograph by Leslie Payne.*

Tynietoy Colonial Mansion with garden. This house unfurnished was priced at $170 in 1930. It came with electric lights and included nine rooms, three halls and a pantry (Not enough examples to determine a price). 2'8" high x 6'2" wide x 1'5.5" deep. *From the collection of Anne B. Timpson. Photograph by Nick Forder.*

The New Model Tynietoy house contains six rooms and two verandas (Not enough examples to determine a price). 27" high x 48" wide x 18" deep. *House and photograph from the collection of Anne B. Timpson.*

Accessories sold by Tynietoy included: clocks, mirrors, lamps, silhouette pictures, maps of Bermuda and also of Nantucket, kerosene heaters, knife boxes, pewter sconces, pitcher, iron, folding screen, shaving mirror on stand, brass candlesticks, hearth brush, tin tray, bread basket, firescreen, and pictures of a windmill or flowers.

The company also sold several different designs of dollhouses for their furniture. See *American Dollhouses and Furniture From the 20th Century* for more information about Tynietoy furniture and houses.

Tynietoy garden which came with the Colonial Mansion. *Garden and photograph from the collection of Anne B. Timpson.*

The living room in the Tynietoy Colonial Mansion included a painted mural. Various pieces of Tynietoy furniture were used to furnish the room. *From the collection of Anne B. Timpson. Photograph by Nick Forder.*

160 Other Wood Dollhouses and Furniture

The bedroom in the Mansion is also furnished with Tynietoy furniture including a four poster bed with canopy. *From the collection of Anne B. Timpson. Photograph by Nick Forder.*

Tynietoy wood kitchen furniture with the original box. Tynietoy was produced from approximately 1920 until the 1950s. The furniture is a large 1" to one foot in scale (Boxed set $300-350). *Furniture from the collection of Gail and Ray Carey. Photograph by Gail Carey.*

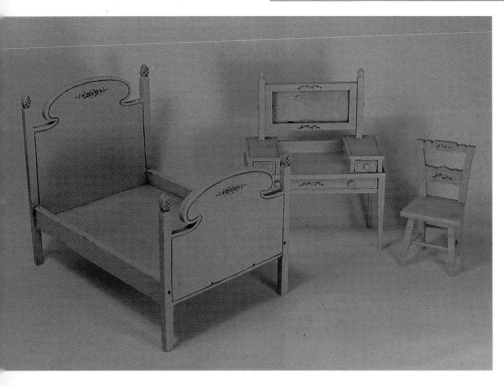

Tynietoy bed with pineapple tops on the posters, matching vanity, and rocker ($225-245 set). *Furniture and photograph from the collection of Patty Cooper.*

Tynietoy living room pieces including fireplace ($75-100), Astor piano ($200-250), Hepplewhyte sofa ($140-160), tilt table ($75-85), lamp table ($50-60), and accessories ($55-75 each). *Furniture and photograph from the collection of Kathy Garner.*

Tynietoy dining room pieces including drop leaf table ($90-100), Chippendale chairs ($50-55 each), and corner cabinets ($100-120 each). *Furniture and photograph from the collection of Kathy Garner.*

Tynietoy grand piano with music box ($250 and up). *Piano from the collection of Gail and Ray Carey. Photograph by Gail Carey.*

Other Wood Dollhouses and Furniture

Miscellaneous Houses

Besides the known companies that made hardboard, Gypsum, Masonite, and wood houses, many other firms competed for the dollhouse market. These lesser known companies are hard to identify. Some of their houses were pictured in the national mail order catalogs or were marked with a trade name but the manufacturer's name was not listed. As the number of dollhouse collectors continues to grow, perhaps their interest will lead to identification of many of these houses.

This house looks very much like one pictured in 1945 advertisements from the Jayline Manufacturing Co. located in Philadelphia ($100-125). The house features scored windows. *House and photograph from the collection of Linda Boltrek.*

This hardboard cottage has a decal on the top of the chimney which reads "Ideal Home/Trademark." The "Ideal" mark is not in the style of the Ideal Toy Co. and the maker is unknown. The heavy plastic windows feature a black diamond pattern like the windows on the early Rich houses ($65-75). 15.25" high x 21.5" wide x 10" deep. *House and photograph from the collection of Marilyn Pittman.*

This fiberboard and wood house is made very much like those made by Schoenhut. It has a wood base, embossed fiberboard siding, and windows that are similar to those used on the Schoenhut Coloniel houses. The windows on this house, however, are made of metal instead of fiberboard. The houses surely must date from the same era, late 1920s, early 1930s. The maker of this house is unknown. The top of chimney has been replaced ($300-350). 18" tall (to roof top) x 19.5" wide x 14.5" deep. *House from the collection of Dian Zillner. Photograph by Suzanne Silverthorn.*

This Tekwood house was pictured in the Sears Christmas catalog for 1948. It sold for $4.69 and featured five rooms on a turn table. It was decorated on the inside and included an awning over the patio ($75-85). 20.5" high x 31.5" wide x 11.75" deep. *House and photograph from the collection of Rita Goranson.*

Hardboard house by unknown maker. The front slides off to give access to the four rooms. Rugs are printed on the floors but the walls are plain ($75-100). 16" to roof peak x 21.25" wide x 11" deep. *House from the collection of Becky Norris. Photograph by Don Norris.*

MISCELLANEOUS WOOD DOLLHOUSE FURNITURE

Barbara Jean

Dollhouse furniture called "Barbara Jean" was produced by Keggs, located in Huntington, Indiana, for a short time circa 1950s. The company made furniture in two different scales: 3/4" to one foot and 1" to the foot. Most of the pieces were finished with flocking. The living room set contained seven items. Since it was given the number 100, it is likely that the firm produced more sets of furniture for different rooms.

"Barbara Jean Doll House Furniture" was manufactured by Keggs located in Huntington, Indiana. The seven piece living room suite No. 103 is in the 3/4" to one foot scale. Circa 1950s (Boxed set $50-55). *Furniture and photograph from the collection of Patty Cooper.*

Barbara Jean furniture was also made in the 1" to one foot scale. Pictured is the seven piece living room suite No. 100 in this size (Boxed set $50-55). The furniture is flocked over wood. *Furniture and photograph from the collection of Patty Cooper.*

164 Other Wood Dollhouses and Furniture

Donna Lee (Woodburn Mfg. Co.)

The wood dollhouse furniture known by the trade name, "Donna Lee" was made by the Woodburn Mfg. Co. located in Chicago. The 3/4" to one foot scale furniture came in boxed sets to furnish the bedroom, bathroom, living room, kitchen, and dining room. The furniture looks very much like the furniture that was sold under the "Nancy Forbes" trade name. Both were probably made during the mid-1940s. The Donna Lee furniture looked cheaper and cost less than the Nancy Forbes products. At least two different boxes were used for the Donna Lee furniture. The earlier box was finished in orange and brown while the newer box style was pink and blue. The earlier boxes did not feature a set of bathroom furniture while the later boxes included a bathroom set. It is known that two different styles of kitchen appliances were produced so it may be that the company revamped all the furniture styles for the newer line.

A dollhouse was also marketed using the Donna Lee name in 1944. It was advertised in the Spiegel catalog at a cost of $2.98 for the four-room furnished house. See *American Dollhouses and Furniture From the 20th Century* for more information on this company.

Boxed Donna Lee living room furniture was made in the mid-1940s by the Woodburn Mfg. Co. in Chicago. It is in the 3/4" to one foot scale. Furniture was also produced for the bedroom, bathroom, kitchen, and dining room (Boxed set $50-60). *Furniture and photograph from the collection of Patty Cooper.*

ECA Toys Inc.

ECA Toys Inc. was in business during World War II. The firm produced at least two different sets of dollhouse furniture. An advertisement from 1944 pictured boxed sets of kitchen and living room furniture. It was made of a composition material and came with paints to decorate the pieces. The living room included sofa, chairs, piano, bench, and console table. The kitchen was equipped with a sink, refrigerator, stove, table, and two stools. The furniture pictured on the box front was beautifully decorated with stripes and other accents. In reality, the furniture came in plain white and a card of watercolors was included to make the transition to the pictured finished product.

Unfinished composition furniture circa 1944 was made by ECA Toys Inc. The same firm also marketed a set of kitchen furniture that was to be finished by the consumer (Boxed set $50). *Furniture and photograph from the collection of Ruth Petros.*

Grand Rapids Furniture
See Wanner

Happy Hour Dollhouse Furniture
See Jaymar

Harco

Harco wood furniture was made in Chicago and each piece was marked "Harco/U.S.A" with a burned-in design. The furniture was produced in at least two different scales. The smaller scale was a small 1" to one foot while the larger items were approximately 1 1/4" to one foot in size. The furniture was first produced by Robert Harper in the 1930s. During the 1960s, Harper made the furniture again using the same patterns he had used earlier. This later furniture was marked exactly like the first editions so it is nearly impossible to distinguish the earlier pieces from those produced in the 1960s. The later furniture was sold through Miniature Mart catalogs issued by John Blauer in San Francisco. The drawers on many pieces were functional. The finish on most of the furniture was a dark walnut with no gloss but some pieces were made with no stain or varnish. The known marked Harco furniture includes desks, bookcases, tables, dining room chairs, rockers, beds, dressers, chests, and magazine rack stand tables.

Wood Harco bedroom furniture was left in the natural finish. This furniture is 1 1/4" to one foot in scale ($15-20 each). *Furniture from the collection of Zelma Fink. Photograph by Suzanne Silverthorn.*

Various pieces of Harco furniture ranged in size from a large 3/4" to one foot scale to 1 1/4" to one foot. Most of the pieces are marked "Harco/U.S.A." ($10-15 each). *Furniture from the collection of Dian Zillner. Photograph by Suzanne Silverthorn.*

Japan

For many decades manufacturers in Japan could produce dollhouse furniture more cheaply than anyone else in the world. Because of this skill, many different kinds of wood dollhouse furniture were imported into various countries to compete with furniture made by local companies. World War II, of course, put a stop to these imports for several years but both before the war and after, Japan remained a high volume dollhouse furniture manufacturer.

Catalogs from the mid 1930s feature several different styles of Japanese furniture. Probably the most common was the small furniture made of wood, reed, and cloth. Several different pieces and sets of this furniture were produced. Included were beds, chairs, sofas, and tables in round and rectangle styles. Most of this furniture was 3/4" to the foot in scale.

Another style of wood furniture often exported from Japan during this same time period was the furniture which featured wood burned designs. Pieces included sofas, tables, chairs, beds, and dressers. Although both of these styles of furniture were made very cheaply, the Japanese also produced some better quality furniture. A boxed living room set circa mid 1930s contains a grand piano, table, and chair, which all appear to be of better quality than the wicker or wood burned pieces. As the years passed, the

Furniture made of wood, reed, and cloth was imported from Japan during the 1930s. The furniture originally sold for 10 cents a set for the 3/4" to one foot scale ($8-10 each). *Furniture and photograph from the collection of Patty Cooper.*

166 Other Wood Dollhouses and Furniture

Japanese craftsmen were able to copy designs from other sources. Some of these ideas came from museums (as used by Shackman) and others were evidently copied from other companies.

According to a former employee of the Strombecker company, several of the original 1930s Strombecker designs were used for furniture that was placed on the market at a later date. It is thought that the unmarked furniture using the Strombecker patterns probably originated in Japan. Pieces have been found in both the 3/4" and 1" to the foot scales. There are several characteristics which distinguish the originals from the copies. The most obvious is that the 1" to the foot Strombecker walnut furniture is almost always marked with the gold "StromBecker Playthings Genuine Walnut" stamp. The copies have no markings and are made of a coarser-grained wood with a duller, browner finish. The gold markings on the Strombecker radio are much crisper than on the copy. The Strombecker dining room chairs have a slight curve to the back whereas the Japanese copies are made of a flat piece of wood. The Strombecker dining room table has six legs, usually with a stretcher, while the copies have only four legs.

Japanese products were sometimes labeled "Made in Japan" as the law required but often this label was made of paper and has vanished with the passing of time. Old catalogs can offer some help in identifying this furniture as it was sometimes called "imported" in the catalog captions. Although this furniture is not now highly collectible, as the years pass and fewer pieces are available, these items may increase in desirability.

This Japanese wood living room furniture in 3/4" to one foot scale also dates from the 1930s. Since the original box pictures a dollhouse full of furniture, it is likely that several other sets were produced ($40-50). *Furniture from the collection of Ray and Gail Carey. Photograph by Gail Carey.*

Original Strombecker walnut furniture is pictured on the right with copies shown on the left. It is thought that the copies were made in Japan. The furniture is all in the 1" to one foot scale (Strombrecker values for large pieces $15-20 and for chair $8-10, reproductions valued at half the Strombecker prices). *Furniture and photograph from the collection of Patty Cooper.*

Jaymar

Most collectors are familiar with the toys made by Louis Marx and Company, but less is known about the Jaymar Specialty Company, which was also associated with the Marx family. This company was financed by Louis Marx in the 1920s and was headed by his father Jacob with help from his sister Rose. The new Specialty Company also became a successful organization for the Marx family. In order for it not to be in direct competition with the larger Louis Marx firm (which specialized in metal and plastic toys), its products were to be made of either wood or cardboard.

Boxed wood kitchen dollhouse furniture was produced by Jaymar in the 3/4" to one foot scale ($65-75). The furniture is circa 1933. The Jaymar Specialty Company was financed by Louis Marx and was headed by his father, Jacob. *Furniture and photograph from the collection of Patty Cooper.*

Other Wood Dollhouses and Furniture 167

The Jaymar company is still in business producing toy pianos and jigsaw puzzles. It is now owned by Jacob Marx's descendants and is still located in New York City. The Jaymar Specialty Co. produced an interesting set of 3/4" to one foot scale wood dollhouse furniture circa 1933. The furniture was called "Happy Hour Doll House Furniture" and five different rooms of furniture were manufactured. It had an "Art Deco" look but the furniture was quite simple. Although the pieces had no functioning drawers or turned legs, the black stenciled designs used on the living room, kitchen, dining room, and bedroom furniture provided an interesting 1930s touch to the finished product. The bathroom fixtures were given a modern look but did not include the stenciling found on the other rooms of furniture.

The Happy Hour line must have been priced very inexpensively as the Montgomery Ward Company offered a cardboard dollhouse and five rooms of the Jaymar furniture for $1.89 in their 1933 Christmas catalog.

Happy Hour pieces made by Jaymar for the living room have an especially appealing "Art Deco" look (Boxed set $65-70). *Furniture and photograph from the collection of Patty Cooper.*

Bathroom furniture from the Jaymar line included towel racks as well as the more usual pieces (Boxed set $65-75). *Furniture and photograph from the collection of Patty Cooper.*

The Happy Hour dining room furniture was decorated with black stenciled designs (Boxed set $65-75). *Furniture and photograph from the collection of Patty Cooper.*

The boxed Happy Hour bedroom furniture included a lamp and candlesticks as well as the basic bedroom pieces (Boxed set $65-75). *Furniture and photograph from the collection of Patty Cooper.*

168 Other Wood Dollhouses and Furniture

The Jaymar Happy Hour furniture was used to furnish a house made of book fiberboard in the Montgomery Ward Christmas catalog in 1933. The house and 36 pieces of the furniture sold for $1.89. *Catalog from the collection of Marge Meisinger. Photograph by Suzanne Silverthorn.*

Kage Co.

The Kage Co., based in Manchester, Connecticut, produced wood 3/4" to one foot dollhouse furniture beginning in 1938 and ending in 1948. Hyman Gerstein was the founder of this company which made room sets of living room, dining room, bedroom, and kitchen furniture. No bathroom pieces were produced. Accessories such as lamps and clocks as well as fireplaces and pianos were also part of the line.

Much of the furniture was upholstered in small print fabrics. The wood dining room, kitchen, and bedroom pieces were painted or scored to indicate drawers but the drawers did not function. The mirrors on the furniture were metal. At first the furniture legs were dowels but later turned legs were used. Later still, a kitchen table and chairs were produced with wire legs attached to wood seats, backs, and table tops. Kibbe Gerstein, a son of the original owner, thinks the legs on these pieces ended in a circular design. The company no longer has any original catalogs as a fire in 1948 destroyed these records. The furniture was sold either in room sets or by the piece for as little as ten cents each. The furniture was marketed mostly through dime stores.

After the end of World War II, the company continued to produce the small furniture for a short time. With the coming of the new plastic dollhouse furniture, Kage could not compete and they discontinued production in 1948. Although the company is still in business, they now produce plastic wall plaques for various holidays. Hyman Gerstein died in 1952 and his son, Kibbe Gerstein, took over the business.

The box lid pictured several sets of 3/4" to one foot furniture made by the Kage Company based in Manchester, Connecticut. The company produced wood furniture for the living room, dining room, bedroom, and kitchen but did not make pieces for the bathroom. *Box courtesy of Roy Sheckells, photograph by Patty Cooper.*

Other Wood Dollhouses and Furniture 169

Kage made a wide variety of living room furniture. The upholstered pieces are especially nice (Larger pieces $10-15, smaller $5-8). *Furniture and photograph from the collection of Patty Cooper.*

The Kage kitchen furniture changed over the years. Chairs, sinks, and tables were first made with dowel legs, and later were upgraded to turned legs (Larger items $10-12, chairs $5-8). *Furniture and photograph from the collection of Patty Cooper.*

Dining room furniture was also made first with dowel legs and later with turned legs (Larger items $10-15, chairs $5-8). *Furniture and photograph from the collection of Patty Cooper.*

Kage beds and vanity dressers included bedspreads and skirts to add interest to the furniture. These dressers and vanities have unusual rectangle metal mirrors (Larger items $10-15, smaller pieces $8-10). *Furniture and photograph from the collection of Patty Cooper.*

Other Wood Dollhouses and Furniture

The more common Kage bedroom pieces had round metal mirrors ($10-15 each). *Furniture and photograph from the collection of Patty Cooper.*

It is known that Kage also produced a kitchen table and chairs with wire legs attached to wood. This set was found in a house that was furnished with other Kage furniture. The table seems to be 3/4" to one foot in scale while the chairs are in the 1" scale (Set $35). *Furniture and photograph from the collection of Lois L. Freeman.*

Keggs
See Barbara Jean

Lincoln (Wright, J.L.)
The most popular product of J. L. Wright, Inc. (based in Chicago) was the famous set of Lincoln Logs that was produced for decades. The logs were designed by John Lloyd Wright in 1916. John was the son of the famous architect Frank Lloyd Wright. The logs were manufactured by the firm until the company was merged with Playskool in 1943. The Wright company also produced other products including dollhouse furniture circa 1936 and 1937. Although the furniture was mainly made of wood, it included steel wire pieces which were used for legs and trim. The furniture was quite modern in design and was approximately 3/4" to one foot in scale. Pieces were made to furnish a bedroom, living room, dining room, kitchen, and sun room. In 1937 the furniture was redesigned and made of a new plastic material instead of wood.

The Lincoln furniture was sold through Butler Brothers catalog in 1936. The catalog described living room furniture (8 pieces), a dining room suite (10 pieces), kitchen furniture (6 pieces), and a bedroom suite (9 pieces). *Photograph by Suzanne Silverthorn.*

Lincoln furniture for the living room included a sofa, chair, ottoman, and radio ($15-20 each) as well as a grand piano, bench, floor lamp, and table not shown. *Furniture from the collection of Becky Norris. Photograph by Don Norris.*

Other Wood Dollhouses and Furniture 171

The kitchen set of Lincoln furniture included six pieces. Besides the stove, sink, and refrigerator pictured ($15-20 each), it is assumed that the other three items were a table and two chairs. *Furniture and photograph from the collection of Rita Goranson.*

Mark Farmer, Inc.

Many pieces of dollhouse furniture were sold through advertisements and catalogs from Mark Farmer, Inc. The firm was located in El Cerrito, California, and the furniture was similar in cost and quality to that sold by Shackman. The furniture was in the 1" to one foot scale. The decades of the 1960s and 1970s were especially profitable for the company. One of their catalogs from 1968 featured dollhouse dolls, country furniture, rugs, accessories (including pictures, lamps, mirrors, rolling pins, tea sets), pianos, organs, and country store items all in the 1" to one foot scale. The furniture was priced from around $4.00 to $15 for each piece. The legs were turned on much of the furniture and the drawers were functional. The firm also sold reproduction china head dolls as well as other products.

This bedroom furniture was sold by Mark Farmer, Inc. in 1962 (Large pieces $20-25, smaller items $10-15). The furniture is in the 1" to one foot scale and featured drawers which functioned. *Furniture and photograph from the collection of Roy Specht.*

This dining room furniture was also sold through Mark Farmer, Inc. in El Cerrito, California, in 1962. Included are a hutch ($25-30), table ($25-30), chairs ($20-25 each), and smaller items ($10-15 each). *Furniture and photograph from the collection of Roy Specht.*

Mary Frances Line

The Victory Toy Company, based in Chicago, produced dollhouse furniture called "The Mary Frances Line" during the mid-1940s. The furniture was made of quarter inch plywood and was slightly larger than 3/4" to one foot in scale. The pieces looked very much like the larger furniture called "Grand Rapids" by collectors. Both sets of furniture were made of plywood and were held together by brads.

The bedroom set contained a bed, night stand, chest of drawers, cradle, rocking chair, and footstool. The boxed living room consisted of a coffee table, sofa, club chair, footstool, and two end tables. Breakfast room pieces included a table, four chairs, and a high chair. A sideboard may also have been made. Since the furniture was quite sturdy, it should still be in good condition when found today.

Wood Mary Frances furniture produced by the Chicago Victory Toy Company circa 1940s. It is in a large 3/4" to one foot scale. The furniture is similar to the "Grand Rapids" pieces but it is smaller. Furniture was produced for a bedroom, living room, and a breakfast room (Large items $15-20, small $5-8). *Furniture from the collection of Mary Lu Trowbridge. Photograph by Bob Trowbridge.*

Additional Mary Frances furniture includes pieces from both the breakfast room and the living room (Large items $15-18, small $5-8). Missing from the sets is a sofa, additional end table, and coffee table. *Furniture from the collection of Mary Lu Trowbridge. Photograph by Bob Trowbridge.*

Miniaform

Unusual sets of dollhouse furniture using the trade name Miniaform were produced in Chicago beginning in 1939. Six rooms of furniture were manufactured: living room, dining room, kitchen, two bedrooms, and bathroom. The pieces were produced by the Hugh Specialty Company and the original boxes were marked "Patent Applied For/Made in USA." The boxed living room set originally sold for 49 cents. The scale of the furniture was 3/4" to one foot.

The furniture was constructed from both wood and wire to give it a modern look. The radio was designed to reflect the new "floor" models. The furniture must not have been on the market long as it is difficult to locate.

Wood and wire furniture in 3/4" to one foot scale was produced by the Hugh Specialty Company in Chicago beginning in 1939. The trade name was Miniaform. The living room pieces came in two different boxed sets. The red box included two living room chairs while the blue box featured a radio (Boxed sets $65-75). *Furniture from the collection of Jill H. Ramsey. Photograph by Gail Carey.*

Other Wood Dollhouses and Furniture 173

This Miniaform bedroom furniture originally sold for 49 cents. The set included two beds, vanity and bench, night table, and chair. According to advertising, two bedroom sets were produced (Boxed sets $65-75). *Furniture and photograph from the collection of Patty Cooper.*

Miniaform kitchen furniture included a table and chairs ($20), refrigerator ($15), stove ($15), and cabinet ($15). The furniture is similar to the Lincoln furniture made by the J. L. Wright Co. *Furniture from the collection of Becky Norris. Photograph by Don Norris.*

Miniaform boxed dining room furniture. The table is missing (Large pieces $12-15, small $8). Furniture was also produced for a bathroom. *Furniture and photograph from the collection of Patty Cooper.*

Nancy Forbes (Rapaport Bros.)

The Rapaport Bros. firm was located in Chicago and they specialized in making wood dollhouse furniture during the 1940s. Most of the company's furniture was in the 3/4" to one foot scale which came in at least two different designs during the years. The firm used the trade name Nancy Forbes for its products. The earlier set was advertised in a catalog from the Famous Funn Family Service Massey's Drug Store, Shirley, Indiana, in 1940. It included the following pieces: Bedroom: bed, blanket chest, night stand, chest of drawers, vanity with mirror, bench and dresser with mirror, and two lamps. Dining room: table, four chairs, buffet with mirror, stacked storage piece, and two lamps. Living room: sofa, fireplace, radio, chair, footstool, coffee table, end table, lamp table, side table, and two lamps. Kitchen: sink, stove, refrigerator, table, four chairs, iron, and ironing board. Bathroom: toilet, lavatory, medicine cabinet, bathtub, scale, hamper, vanity with mirror, and bench. Child's room: youth bed, blanket chest, steps, night stand, chifforobe, two lamps, and a small table and chair. One room of furniture sold for 98 cents in 1940.

Apparently much of this furniture was also made in the 1" to one foot scale during the early 1940s. Although the pieces are similar, the larger furniture included functioning drawers and more detailed designs. It is much harder to locate for today's collectors.

By the mid 1940s a new line of furniture had been designed by the company and it was featured in the catalogs of that period (see *American Dollhouses and Furniture From the 20th Century* for more information).

Left: Early Nancy Forbes living room furniture circa 1940 (Large pieces $12-15, small $4-6). The pieces are 3/4" to one foot in scale. *Furniture and photograph from the collection of Patty Cooper.*

Below left: The early Nancy Forbes bedroom was furnished with a bed, blanket chest, vanity and bench, dresser with mirror, chest of drawers, nightstand, and lamps (Large items $12-15, small $4-6). *Furniture and photograph from the collection of Patty Cooper.*

Below right: The same line of Nancy Forbes furniture included seven dining pieces plus lamps. One of the chairs is not pictured (Large pieces $12-15, small $4-6). *Furniture and photograph from the collection of Patty Cooper.*

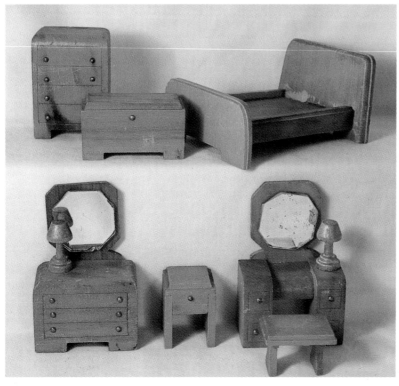

Even the Nancy Forbes bathroom included extra pieces: a medicine cabinet, vanity and stool, scale, and a hamper (missing in photograph) (Large pieces $12-15, small $4-6). *Furniture and photograph from the collection of Patty Cooper.*

The Nancy Forbes nursery furniture was similar to the Strombecker 1" to one foot nursery but the doors and drawers did not function on any of the Forbes pieces. The nursery furnishings also included a small table and chair (Large items $12-15, small $4-6). *Furniture and photograph from the collection of Patty Cooper.*

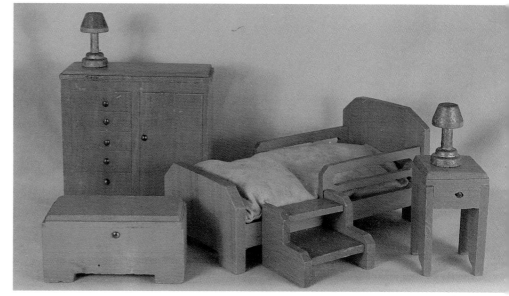

The Forbes kitchen included the pieces shown plus three more chairs, and an ironing board with iron (Large items $12-15, small $4-6). *Furniture and photograph from the collection of Patty Cooper.*

Larger 1" to one foot Nancy Forbes wood furniture was also produced. The drawers on this furniture were functional. The bedroom included a bed, nightstand, chest of drawers, blanket chest, vanity, and dresser with mirror (Large items $18-25, small $10-12). *Furniture from the collection of Marci Tubbs. Photograph by Bob Tubbs.*

The larger Nancy Forbes dining room pieces included a table, four chairs, server, buffet, and china cabinet (Large pieces $18-25, small $10-12). *Furniture from the collection of Marci Tubbs. Photograph by Bob Tubbs.*

Queen Anne (Packman, B.)

B. Packman, a firm located in New York circa 1940s, produced some unique dollhouse furniture using the trade name, "Queen Anne Upholstered Doll House Furniture." The furniture was a large 3/4" to one foot in scale and combined pieces upholstered in rayon or taffeta floral prints over wood with furniture made of clear Lucite.

At least three boxes of furniture were made. The living room included a sofa, two chairs, two plastic tables, table lamp, and floor lamp. The bedroom pieces consisted of two beds, a chair, dressing table, plastic night stand, and a radio. The other set of furniture appeared to be for a dressing room or sitting room. The items for this room included a chaise lounge, sectional love seat, floor lamp, two clear plastic tables, table lamp, and radio. Fringe was glued around the bottom of each piece of upholstered furniture.

An original box pictures the "Queen Anne Upholstered Doll House Furniture." The furniture was made in a large 3/4" to one foot scale and included sachet. B. Packman in New York produced the pieces. *Box and photograph from the collection of Lois L. Freeman.*

Other Wood Dollhouses and Furniture 177

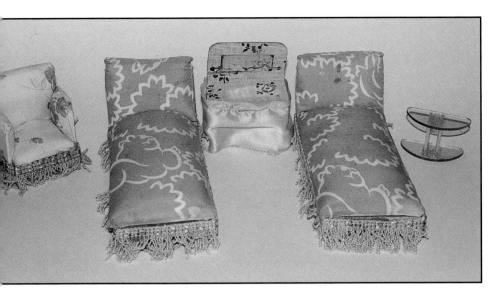

Most of the Queen Anne furniture was upholstered in rayon or taffeta floral print over wood while some of the pieces were made of clear Lucite. The bedroom included two chaise lounges, a vanity, chair, nightstand, and a radio (Boxed set $75-85). *Furniture and photograph from the collection of Lois L. Freeman.*

Queen Anne pieces from the living room set ($15-18 each). Also included in the living room were two plastic tables, table lamp, and a floor lamp. *Furniture and photograph from the collection of Lois L. Freeman.*

Another boxed set of Queen Anne furniture included a chaise lounge, a two-piece sectional, floor lamp, two clear plastic tables, table lamp, and a radio ($15-18 each). *Furniture and photograph from the collection of Lois L. Freeman.*

Rapaport Brothers
See Nancy Forbes

Shackman, B. & Co.

B. Shackman & Co. was a wholesale importing firm located in New York City. Although the company was founded in 1898, it is the firm's dollhouse furniture dating from the 1970s that is most often found by today's collectors. Their catalog listed more than 200 items for the dollhouse enthusiast during that time. This included both 1" to one foot scaled furniture and accessories. Most of their items were made in Japan and the furniture represented all styles including Sheraton, Duncan Phyfe, and Early American. The furniture was moderately priced so dollhouses for both children and adults were furnished with Shackman items during the 1970s. The new kit houses were just beginning to appear on the market and the Shackman furniture was used by many of these hobbyists to furnish dollhouses. Shackman also sold reproduction bisque dolls that could be used in dollhouses.

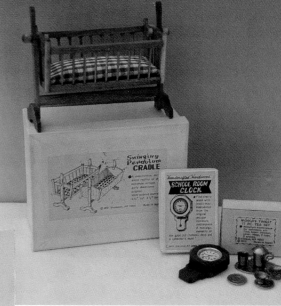

Shackman's catalog from the 1970s listed more than 200 dollhouse items for sale including dolls and accessories ($15-20 per box). *From the collection of Gail and Ray Carey. Photograph by Gail Carey.*

Furniture sold by B. Shackman and Co. during the 1970s ($15-20 per box). The furniture is 1" to one foot in scale. The boxes carry the Shackman, N.Y., name but also state that the furniture was Made in Japan. The company also sold other products made in Germany and England. *Furniture from the collection of Dian Zillner. Photograph by Suzanne Silverthorn.*

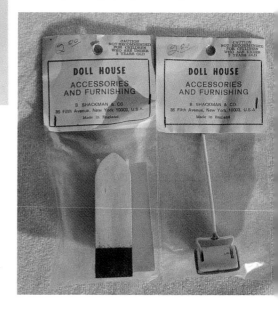

These accessories were also sold by Shackman but unlike most of their products, this ironing board and carpet sweeper were made in England ($10-12 each). The sweeper is identical to the one marketed by Barton. *From the collection of Gail and Ray Carey. Photograph by Gail Carey.*

Star Novelty Works

Star Novelty Works was located in Cincinnati, Ohio, in the early part of the twentieth century. The firm's dollhouse furniture has been identified as being sold through an R. H. Macy 1910-1911 catalog by Flora Gill Jacobs in her book *Dolls' Houses in America*. The boxed parlor set in the advertisement included three straight chairs, one chair with arms, a rocker, a settee, a round table, and a large china cabinet. The set sold for 98 cents in the catalog. The furniture was approximately 1 1/4" to one foot in scale. A boxed Star oak dining room set circa 1910 is a little larger in scale being 1 1/2" to one foot. The box is marked "American Toy Furniture/Manufactured by Star Novelty Works, Cincinnati, Ohio U.S.A./One Dining Room Set No. 302." It is likely that the company produced other rooms of furniture as well.

Boxed set of 1 1/2" to one foot scale oak dining room furniture carries the label of the Star Novelty Works from Cincinnati, Ohio. The furniture is circa 1910-1915. *Furniture from the collection of Dian Zillner. Photograph by Suzanne Silverthorn.*

Other Wood Dollhouses and Furniture 179

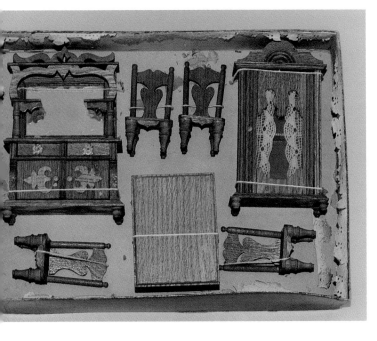

This oak bedroom furniture in the 1 1/4" to one foot scale is also attributed to Star Novelty Works. The pulls are identical to ones on other Star pieces. The drawers are functional ($175-225 set). *Furniture from the collection of Dian Zillner. Photograph by Suzanne Silverthorn.*

The boxed Star dining room includes a table, four chairs, a buffet, and a china cabinet (Boxed $275-300). The set is similar to the dining room furniture pictured in the section on Cass. The Star chairs have straight backs instead of curved. *Furniture from the collection of Dian Zillner. Photograph by Suzanne Silverthorn.*

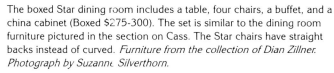

Star also produced a parlor set which included two straight chairs, a rocker, settee, table, and curio cabinet. According to Flora Gill Jacobs, a similar set was sold through R. H. Macy in 1910-1911 for 98 cents. The furniture is 1 1/4" to one foot in scale ($175-225 set). *Furniture from the collection of Dian Zillner. Photograph by Suzanne Silverthorn.*

Tinker Toys (The Toy Tinkers, Inc.)

The Toy Tinkers, Inc. was incorporated in 1923 in Evanston, Illinois. The business had been started in 1914 when Charles Pajeau patented and began producing the famous Tinker Toys. The factory also manufactured other games and toys through the years.

Of interest to dollhouse collectors is the set of wood furniture pictured in the company catalog for 1931. Apparently only one room of furniture (living room) was produced. The product must not have met with much success as it was soon discontinued. The set included thirteen pieces and was finished in walnut.

A. G. Spaulding and Bros., Inc. purchased the company shortly before Pajeau's death in 1952. Spaulding continued to operate the business in Evanston until the 1970s when it was purchased by Questor Education Prod. Co., a New York manufacturer. Gaberiel Industries acquired the company in 1978.

Tinker living room furniture in the 3/4" to one foot scale. The set included thirteen pieces and was finished in walnut with colorful accents ($12-15 each). *Furniture from the collection of Mary Lu Trowbridge. Photograph by Bob Trowbridge.*

180 Other Wood Dollhouses and Furniture

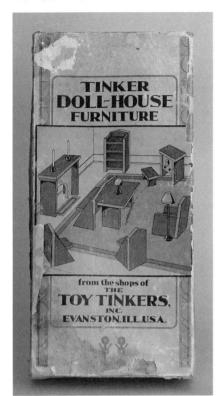

Box cover for wood furniture made by The Toy Tinkers in Evanston, Illinois circa 1931 (Boxed set $135-150). *Box from the collection of Mary Lu Trowbridge. Photograph by Bob Trowbridge.*

Additional pieces of Tinker furniture could have been used in a dining room as well as a living room ($12-15 each). *Furniture from the collection of Mary Lu Trowbridge. Photograph by Bob Trowbridge.*

Toncoss Miniatures

Toncoss Miniatures was a line of 1" to one foot scale dollhouse furniture fashioned in the Early American style. The firm that produced the furniture was headed by Ewart W. Tonner of Sturbridge, Massachusetts, and Alfred A. Habercoss of Connecticut who was Tonner's son-in-law. The furniture name was a combination of the partner's names. In 1958 the furniture was made by Habercoss. The firm sold the same designs of furniture into the 1970s with the addition of two extra items.

The pine furniture included the following pieces: stretcher table, kitchen chairs, dry sink, Dutch cupboard, rocker, wing chair, several designs of fireplaces, Governor Winthrop desk, four poster bed with canopy, highboy, lowboy, blanket chest, candle stand, commode, four drawer chest, night stand, wash stand, trundle bed, hutch table, drop leaf table, cradle, and cobbler's bench. Each piece of furniture is marked on the bottom "Toncoss/Sturbridge" so it is easy to identify.

This bedroom furniture was made by Toncoss Miniatures circa 1970s. The Early American styled furniture is in the 1" to one foot scale and was made of pine. Pieces include a canopy bed, blanket chest, highboy, lowboy, chair, fireplace, and wash stand. The furniture is pictured in the Toncoss room that was made to display the furniture (Small items $8-12 each, large $20-30). *Furniture and photograph from the collection of Kathy Garner.*

Toncoss kitchen furniture included a dry sink, trestle table and chairs, large fireplace, Dutch cupboard, and commode. Each piece is marked "Toncoss/Sturbridge." (Small items $8-12 each, large $20-35). *Furniture and photograph from the collection of Kathy Garner.*

Other Wood Dollhouses and Furniture 181

Toncoss furniture that could be used in a child's room include a cradle, trundle bed, four drawer chest, nightstand, and rocking chair (Small items $8-12 each, large $20-35). *Furniture from the collection of Kathy Garner and Zelma Fink. Photograph from the collection of Kathy Garner.*

Toncoss living room furniture includes wing chairs, fireplace, Governor Winthrop desk and chair, candle stand, and cobbler's bench. The pieces are shown in the Toncoss room setting (Small items $8-10 each, large $20-35). *Furniture and photograph from the collection of Kathy Garner.*

Victory Toy Company
See Mary Frances Line

Wanner
The furniture known to collectors as "Grand Rapids" was much larger in scale than the similar Mary Frances pieces being nearly 1 1/2" to one foot in scale. It may have been made by Wanner as several of the pieces are marked inside a circle "Wanner/1933/U.S.A./Since." It is not known if the furniture was made continuously from 1933 or if (like Harco) the pieces were produced again at a later date. Many of the items have been found with black and silver stickers on the bottoms indicating prices of 20 to 25 cents each. The stickers appear to be newer than the 1930s. Since the marked furniture does not carry a city for Wanner, it may be that the firm was also from the Chicago area as so many of the similar companies were. The furniture was sold by the piece in dime stores. Furniture was made to furnish a living room, dining room, and bedroom with extra pieces that could be used in a baby's room or study. The line included the following items: Living room: sofa, matching arm chair, wing chair, oval lamp table, step lamp table, magazine rack, half moon lamp table, and book case or display piece on legs. Bedroom: beds, rectangle night stand, dresser with mirror, chest of drawers, and blanket chest. Dining room: table and chairs (two different designs of chairs were made), server, and hutch. Baby items include a cradle and a rocking chair. An icebox was also produced and could have been used with the table, chairs, and hutch to furnish a kitchen. This furniture was also quite sturdy and is usually found in very good condition. Many of the items were marked "Made in U.S.A." and other pieces had no identification. The furniture that was marked "Wanner" is rare. Although designs and the color of finish differ slightly among pieces of the "Grand Rapids" furniture, the variations are so slight, it seems likely that all of the items were produced based on the patterns from one company.

Wood living room furniture in a scale of approximately 1 1/2" to one foot (Large items $15-20, small $5-8). Some collectors call this furniture "Grand Rapids" but it may have been produced by Wanner. The pieces appear to have been made from plywood and the nails used to assemble the furniture are visible on each item. *Furniture from the collection of Dian Zillner. Photograph by Suzanne Silverthorn.*

182 Other Wood Dollhouses and Furniture

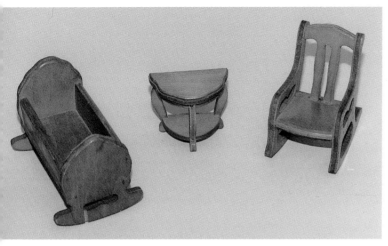

A rocker and a cradle were also produced which could be used in a child's room (Large items $15-20, small $5-8). *Furniture from the collection of Dian Zillner. Photograph by Suzanne Silverthorn.*

Woodburn Mfg. Co.
See Donna Lee

Wright, J.L.
See Lincoln

Miscellaneous Wood Dollhouse Furniture
Other companies also produced wood dollhouse furniture from time to time. Some of this furniture can be found in old mail order catalogs. Although the maker is usually not identified, the time period for the furniture can be established as well as its original cost. Perhaps with more research from old advertisements, the puzzle of the origin of these pieces may be solved.

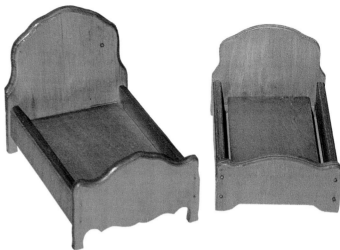

These two beds are very similar but the bed on the right is marked inside a circle "Wanner/1933/U.S.A./Since." The Wanner bed is 3.25" tall x 3.75" wide x 5.75" long. The bed on the left is marked "Made in U.S. of America." It measures 4.5" tall x 3.75" wide x nearly 6" long ($15-20 each). *Furniture from the collection of Becky Norris. Photograph by Don Norris.*

These 1" to one foot nursery pieces were sold in the Chestnut Hill catalog in 1951 along with a room box. Besides the items pictured, other pieces included another chair, lamps, and night table ($25-35 each). *Furniture and photograph from the collection of Lois L. Freeman.*

This wood kitchen set in the 1" to one foot scale was sold from approximately 1950 until the early 1960s ($35-40 each). Although the maker is unknown, it was advertised with a nursery set and later a bathroom. *Kitchen and photograph from the collection of Patty Cooper.*

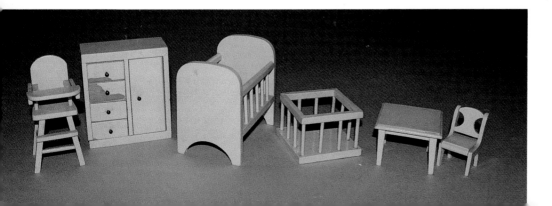

Other Wood Dollhouses and Furniture 183

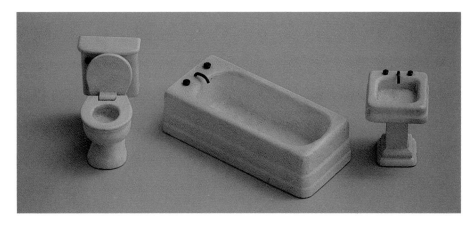

This bathroom set was featured in the 1951 Chestnut Hill catalog and appeared with the wood kitchen furniture in the 1961-1962 Mark Farmer catalog ($50-75). It is in a small 1" to 1 foot scale. The set was priced at $4.00 in 1961. *Furniture from the collection of Dian Zillner. Photograph by Suzanne Silverthorn.*

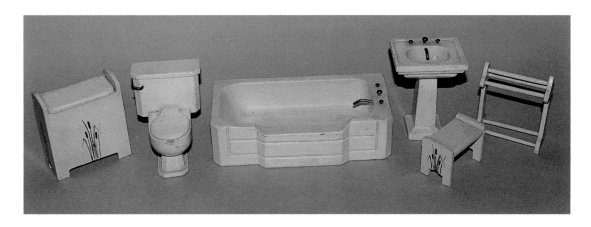

This 1" to one foot scale bathroom set with a handpainted reed design is marked "Made in U.S.A." with red lettering on a white label ($75-100). The maker is unknown. *Furniture and photograph from the collection of Lois L. Freeman.*

Bedroom furniture in a small 1" to one foot scale includes flower designs on each piece. The drawers are functional. Circa 1930 ($125-150 set). *Furniture and photograph from the collection of Patty Cooper.*

These living room pieces also include flower designs and may be made by the same unknown company as the pink bedroom set. These are also in a small 1" to one foot scale (Large items $18-25, small $15-20). *Furniture and photograph from the collection of Patty Cooper.*

Wood kitchen furniture which may have been produced by the same maker features the same type knobs as those used on the pink bedroom pieces. These are also in the 1" to one foot scale (Large items $18-25, small $15-20). *Furniture and photograph from the collection of Patty Cooper.*

METAL FURNITURE

Althof, Bergmann and Co.

Althof, Bergmann and Co. was located at Park Place in New York City. Their catalog from 1874 stated that it included tin and mechanical toys manufactured by Althof, Bergmann and Co. The catalog also mentioned that the firm imported "Toys and Fancy Goods." In addition, china and Bohemian Glassware were included in their line of merchandise. Pictured in the catalog were four rooms of dollhouse furniture. These included sets for the parlor, bedroom, dining room, and kitchen. The furniture was made of painted tin in the Victorian style with the addition of stenciled designs to make the furniture more appealing. The living room pieces included a Victorian sofa and chair, six straight chairs, round table, mirror, corner shelf, and a fireplace. The bedroom furniture included a fireplace, four straight chairs, mirror, washstand, towel rack, bureau, and bed. The dining room furniture consisted of six straight chairs, round table, fireplace, high-backed buffet, and a lounge. The kitchen pieces included four straight chairs, two tables, bench, and a cabinet.

The company began operation around 1867 and used the trademark "ABC." Pictured are sets of furniture as shown in the 1874 catalog.

Tin bedroom pieces pictured in the 1874 catalog of Althof, Bergmann and Co. located in New York City. The furniture was approximately 1 1/4" to one foot scale. Besides the pieces shown, a bureau, two additional chairs, fireplace, mirror, and towel rack were also part of the original set (Bed $400-500, chairs $175-225 each, washstand $250-350). *Furniture from the collection of Gaston and Joan Majeune. Photograph by Renee Majeune.*

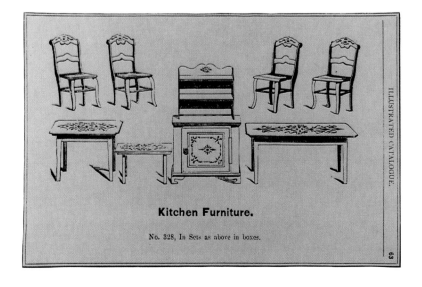

The Althof, Bergmann tin furniture for the kitchen also included stenciled patterns for decoration. These pieces were pictured in the firm's 1874 catalog. *Catalog from the collection of Paige Thornton. Photograph by Suzanne Silverthorn.*

Arcade

The Arcade Company was founded in Freeport, Illinois in 1885. Their iron dollhouse furniture was produced from 1925 to 1936. The house and the individual rooms were made to accommodate the 1 1/2" furniture. There were other scales made as well including 1" to one foot and a smaller scale of approximately 3/4" to the foot. The room settings featured printed curtains, windows, and other wall decorations. A living room, dining room, kitchen, dining alcove, bedroom, bathroom, and laundry room were made. See *American Dollhouses and Furniture From the 20th Century* for more information on Arcade products.

Arcade produced two different sizes of dollhouses to house their iron furniture. Pictured is the larger one (circa 1930s) which is 34" high x 9' 11" wide x 18.5" deep. The house contains nine rooms plus a hall and breakfast nook as well as a garage. It is furnished with Arcade furniture. The cardboard inserts that Arcade produced for their furniture could also be used to decorate the houses (Not enough examples to determine a price). *House from the collection of Gaston and Joan Majeune. Photograph by Renee Majeune.*

Arcade living room furniture included a grand piano ($600 and up), sofa and chair ($750 and up), ladderback chair ($175 and up), secretary ($400 and up), and not pictured a reading table and end table. *Furniture from the collection of Gaston and Joan Majeune. Photograph by Renee Majeune.*

Arcade bedroom furniture included a bed ($300 and up), dresser ($300 and up), chair ($125 and up), desk ($300 and up), and rocker ($150 and up). All of the Arcade furniture was made of iron. *Furniture from the collection of Dian Zillner. Photograph by Suzanne Silverthorn.*

Metal Furniture 187

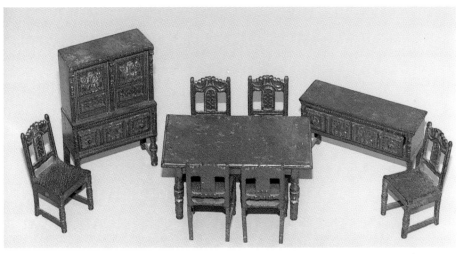

Arcade dining room pieces included a table ($250 and up), six chairs ($100 and up), a buffet, and china closet ($250 and up each). The doors and drawers were functional. *Furniture from the collection of Dian Zillner. Photograph by Suzanne Silverthorn.*

Arcade also made cardboard rooms for each set of furniture. The laundry room included a washer ($300 and up complete), ironer ($250 and up), laundry tubs ($250 and up), chair ($75 and up), laundry tray ($75 and up), heater ($100 and up), and boiler ($75 and up). The room measures 10.25" high x 14" wide x 11" deep ($200 and up). *Furniture from the collection of Dian Zillner. Photograph by Suzanne Silverthorn.*

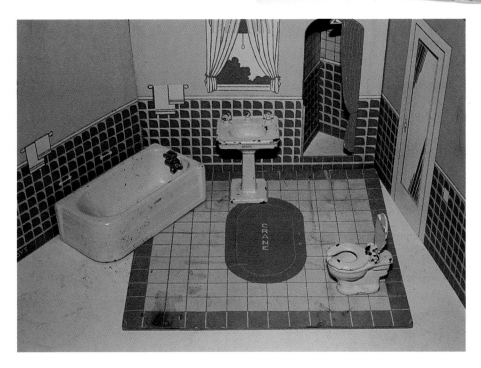

The Arcade iron bathroom furniture included a bathtub, toilet, lavatory, and shower stall to furnish the cardboard bathroom. A stool was also included in the set. All of the pieces to be used in the houses were scaled at 1.25" to one foot (Fixtures $125 and up each). *Furniture from the collection of Joan and Gaston Majeune. Photograph by Renee Majeune.*

188 Metal Furniture

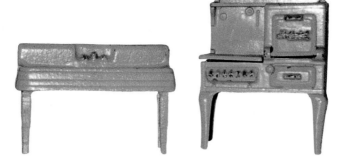

Besides the large iron pieces, Arcade also produced some smaller items of furniture. Pictured are three kitchen appliances approximately 1" to one foot in scale ($100-125 each). *Furniture from the collection of Sally Hofelt. Photograph by Kathy Garner.*

This iron Arcade bathroom furniture is even smaller being approximately 3/4" to one foot in scale ($75-100 each fixture, stool $40-50). *Furniture and photograph from the collection of Patty Cooper.*

Brown, George W. and Co.

George W. Brown and Co. was founded in Forestville, Connecticut, in 1856 and remained in business until 1868. The firm produced tin toys which included sets of dollhouse furniture for the parlor (made in imitation rosewood) and bedroom (with an oak grained look). The firm also made a complete tin kitchen. The pressed tin parlor set included a sofa, table, four armless chairs, one chair with arms, and two ottomans. The tin bedroom furniture consisted of a mirrored dresser, two chairs, washstand, bed, towel rack, and table. Both of these sets of furniture were very Victorian in style.

In 1868 the company was sold and in 1869 George Brown joined with Elisha Stevens of J. & E. Stevens to form the Stevens & Brown Manufacturing Co. in Cromwell, Connecticut. The two men opened a sales room in New York with the trade name American Toy Company. Brown and Stevens issued a catalog under the name Stevens & Brown Manufacturing Co. in 1872. Since the Stevens firm produced iron toys, and Brown issued products made of tin, the sales merger offered consumers a wider range of toys. The partnership ended in 1880.

These two tin Victorian chairs ($200-25 each), ottoman ($80-110), and clock ($175-200), were pictured in the 1872 catalog of the Stevens & Brown Manufacturing Co. of Cromwell, Connecticut. Since the George Brown firm produced tin items while Stevens was known for iron toys, it is assumed this furniture was made by Brown. The other tin chair ($175-225) was also pictured in the 1872 catalog as part of a bedroom set. The furniture is approximately 1" to one foot scale. The stove is unidentified. *Furniture from the collection of Gaston and Joan Majeune. Photograph by Renee Majeune.*

Cooke, Adrian

The Adrian Cooke Metallic Works was located in Chicago and the company produced furniture similar to that marketed by Peter Pia circa 1895-1905. Their pieces were advertised as being made of an alloy of aluminum and white metal. There were at least three different weights of this product used to produce several different sets of furniture. The company made many designs of furniture in several different scales ranging from 1/2" to one foot to 1" to the foot. The earlier items were marked "Patent Applied For" while the later furniture was marked with a patent date. Some of the company labels refer to the toys as "Fairy Furniture."

The patent date stamped on many of the larger pieces of furniture is August 13, 1895. This patent was issued to Joseph J. Jones of Brooklyn, New York. The drawing and information listed by the U.S. Patent Office related to the chair and sofa cushions. Instead of using flat seats as did the Peter Pia firm, the Cooke cushions had a concave form with a middle that was higher than the sides. Included in the Cooke line were arm chairs, straight chairs, rocking chairs, sofas, tables, and beds.

Tiny 1/2" to one foot scale metal furniture was made by Adrian Cooke Metallic Works in Chicago circa early 1890s. The furniture came with this card reading "The Fairy Furniture/Indestructible/An Alloy Of/Aluminum and White Metal/Manufactured By/Adrian Cooke Metallic Works/Chicago, Ill./Patent Applied For." The furniture also included "Pat. App. For" incised on the front of each piece. *From the collection of Dian Zillner. Photograph by Suzanne Silverthorn.*

This set of the small Adrian Cooke furniture included beds in addition to the parlor pieces featured in the original set. On this furniture is a patent date stamped onto the furniture ($25-30 each). *Furniture and photograph from the collection of Patty Cooper.*

This larger set of furniture attributed to Cooke includes a settee ($45-50), rocking chair ($35-40), straight chairs ($30-35), and a table ($50-55). The scale is approximately 3/4" to one foot. This same type of furniture was also produced with a heart design on the back. Each piece is stamped with the patent date August 13, 1895. The patent was issued for a chair cushion as shown with a concave form. *Furniture and photograph from the collection of Patty Cooper.*

This larger 1" to one foot scale furniture was made to honor Columbus and is attributed to Cooke. The pieces are stamped with the August 13, 1895, patent date and the cushions have a concave form. The furniture was advertised in the 1895 Butler Brothers wholesale catalog priced at $1.80 for a dozen sets. Another similar set of furniture with an anchor design has been found with the Cooke label. The metal is heavier on this furniture than on the smaller pieces and the chair measures 3.5" high ($150-175 set). *Furniture from the collection of Dian Zillner. Photograph by Suzanne Silverthorn.*

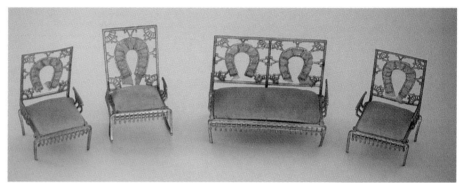

This set of "anchor" furniture was also produced by Adrian Cooke. The pieces are approximately 1" to one foot in scale and have ribbons woven throughout the anchors for decoration ($150-175 set). *From the collection of Dian Zillner. Photograph by Suzanne Silverthorn.*

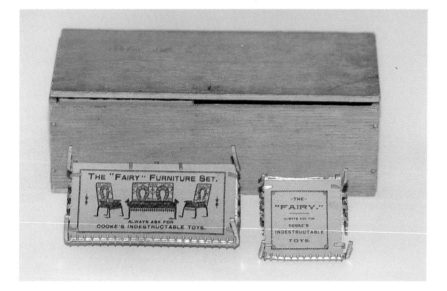

Pictured is the original box for the Cooke furniture as well as the labels from the bottoms of the settee and chair which read "The 'Fairy' Furniture Set/Always ask for Cooke's Indestructible Toys." *From the collection of Dian Zillner. Photograph by Suzanne Silverthorn.*

This Adrian Cooke furniture has the company name incised on the sofa. It says "Cooke Metallic Works Chicago, Ill." The sofa is 4" tall and 4.75" wide. Although the metal is heavier than that used on the other furniture, it still bends quite easily ($150 for the set). *Furniture from the collection of Dian Zillner. Photograph by Suzanne Silverthorn.*

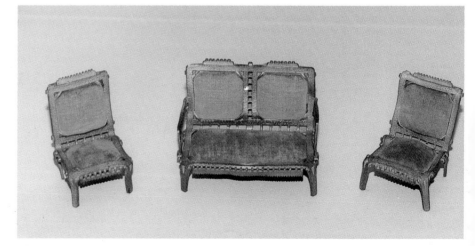

French Metal Furniture

Many of the early dollhouse books contained pictures of furniture and dollhouses which were said to be French. The charm of these pieces, sometimes combined with the location in which they were originally found, made this a likely guess. But more recent researchers have found that most of these pieces were actually made in Germany and very few dollhouse items can be verified as French. However, the "French Penny Toys" were one category of furniture and accessories which were definitely made in France. These metal pieces usually have "France" actually cast into them. Sometimes, they are also marked with the name Simon et Rivollet. It is not known if this company manufactured all of the items known as "penny toys" but they are obviously responsible for a large portion of them. The company was in business from the 1890s to the 1920s.

The furniture was made in a frustrating mixture of sizes with an inadequate variety in any one scale to furnish a house. The beds are tiny, less than 1/2" to one foot in scale, while some of the accessories are compatible with 1" to one foot scaled furnishings. The furniture may be found in a range of colors including green, gold, and silver. The folding stroller marked "S&R" is identical to that shown in the 1925 Tootsietoy catalog, so apparently Dowst imported some pieces from France.

Since much of this dollhouse furniture is so tiny, the pieces can be used to furnish the smallest of the Bliss and "Blue Roof" houses even though the country of origin would not be in keeping with that of the houses.

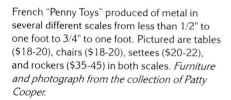

French "Penny Toys" produced of metal in several different scales from less than 1/2" to one foot to 3/4" to one foot. Pictured are tables ($18-20), chairs ($18-20), settees ($20-22), and rockers ($35-45) in both scales. *Furniture and photograph from the collection of Patty Cooper.*

Although the French metal beds were quite small ($40-50), they are quite attractive when displayed with vanities ($30-35) and other bedroom pieces (mirror $45-55). Most of this furniture was made by the Simon et Rivollet company circa 1910-1920. *Furniture and photograph from the collection of Patty Cooper.*

French baby strollers and buggies ($45-50). Most of these items have the word "France" cast into them as do many of the French cast metal pieces. *Furniture and photograph from the collection of Patty Cooper.*

The French metal line also included pianos ($25-30) and kitchen pieces (stoves $40-45) as well as furniture for a nursery, bedroom, dining room, and living room. *Furniture and photograph from the collection of Patty Cooper.*

French baby furniture included high chairs ($60-65), bassinets ($85-100), and strollers ($45-50). A sewing machine was also part of the line ($25-30). *Furniture and photograph from the collection of Patty Cooper.*

192 Metal Furniture

Peter Pia

According to Marshall and Inez McClintock in their book *Toys in America*, Peter Pia began producing pewter toys in New York around 1848. Dollhouse furniture made of pewter alloy was advertised under the Peter F. Pia name in *Playthings* magazine in 1905. Similar furniture had been featured earlier in the 1900 Montgomery Ward mail order catalog where a furnished folding parlor room was offered for sale. The room, complete with furniture, sold for only twenty-five cents. The cardboard room was 12" square and came with a settee, two chairs, table, mirror for the fireplace, framed pictures, and easel. In 1899, E. W. Blatchford and Co. advertised this same "Doll's Folding House." The firm listed a folding dining room which included a table, four chairs with seats covered in imitation leather, mantel, pictures, and easel. Their advertising stated that the furniture was made of "new lustrous and unbreakable metal neatly decorated in gold." A folding bedroom was offered which was furnished with a bed, chair, rocker, mantel, pictures, and easel. The easel also came separately and the firm suggested it could be used as a summer resort souvenir or as a holder for a business Christmas greeting. A kitchen which included a range, table, and two chairs was also produced. Included with the Blatchford listing was an offer from Sanford Brothers (Chicago, Illinois) to enable the consumer to receive any of the rooms free when five boxes of their toilet soap were purchased. Each of the rooms contained printed walls and carpet and the outside walls were imitation brick. The corners of the rooms were doubly hinged with book cloth.

Although this metal furniture is highly collectible, it is hard to really determine a maker. It has been attributed to Peter F. Pia but in the Blatchford advertisement, the furniture is called "our new lustrous and unbreakable metal." Perhaps the company was just indicating that the furniture was being sold by them for the first time in their catalog dated 1899.

Another company that made similar furniture was the Adrian Cooke Metallic Works of Chicago. This company marked many of their pieces of furniture with a paper label which included their name. They used "The 'Fairy' Furniture" as their trade name (see Cooke, Adrian).

It is difficult for today's collectors to sort out the makers of these similar pieces of metal furniture. Each item of marked or boxed furniture that can be identified will help future collectors match their unmarked furniture for a positive identification. The furniture is highly valued by collectors who are trying to find appropriately scaled furniture to fit houses such as those made by Bliss or Dunham's Cocoanut. This furniture is sought for these houses because it is American made and is appropriate in both size and time period.

Stevens, J. & E.

J. & E. Stevens & Co. was founded in Cromwell, Connecticut, in 1843. The company became famous for making iron toys including banks and wheel toys. They also produced many different designs of iron dollhouse furniture of various sizes. Chairs ranged from 2.75" to 5.5" in height and tables were from 1.25" to 4.25" tall. Sofas were sized from 3" to 6.5" in length. Beds ranged in size from 4.75" to 2 feet long. Other items included bureaus, mirrors, and cradles. The Victorian parlor set in a velvet finish included four chairs, sofa, rocking chair, table, and two ottomans. A similar set was also made in a satin finish.

For a brief period from 1868 until 1880 Elisha Stevens combined with George W. Brown in a sales merger to form the Stevens and Brown Manufacturing Company. They issued a combined catalog in 1872 under this new name and opened a New York sales room using the tradename American Toy Company. The partnership ended in 1880. The Stevens firm continued to sell metal toy stoves until the 1930s.

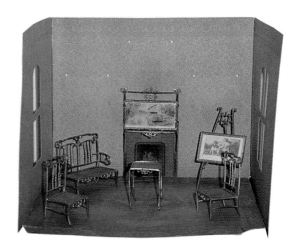

Cardboard furnished parlor attributed to Peter Pia of New York City circa 1900 ($300-350). These rooms were sold by Montgomery Ward through mail order in 1900 and by E.W. Blatchford and Co. in 1899. Other rooms were produced including a dining room, bedroom, and kitchen. The furniture was made of "new lustrous and unbreakable metal neatly decorated in gold." The parlor included a settee, two chairs, table, mirror, framed pictures, and an easel. The room was 12" square. Although the furniture looked very much like that made by Adrian Cooke, the chair seats were flat instead of curved. *Room from the collection of Leslie and Joanne Payne. Photograph by Leslie Payne.*

The bed from the bedroom setting attributed to Peter Pia is pictured here ($125-150). Other items in the furnished bedroom included a chair, rocker, mantel, pictures, and easel. All of these rooms together made up the "Doll's Folding House." *Furniture and photograph from the collection of Patty Cooper.*

J.& E. Stevens Co. of Cromwell, Connecticut produced iron toys for many years. Pictured is a parlor set dating from the 1880s. Included are a sofa 6.5" long ($600-650); chair 4.5" tall ($175-225); table 3" tall ($250-350); and ottomans ($80-110). *Furniture from the collection of Gaston and Joan Majeune. Photograph by Renee Majeune.*

Metal Furniture 193

Stevens dining room table and chairs. The table is 5" long ($400-500) and the chairs are 3.75" tall ($150-165). *Furniture from the collection of Ray and Gail Carey. Photograph by Gail Carey.*

Pictured is a chair 3.75" tall ($175-225) and a table ($300-375) made by J.& E. Stevens Co. circa 1880s. *Furniture from the collection of Gaston and Joan Majeune. Photograph by Renee Majeune.*

Stevens also produced a set of iron bedroom furniture including a bureau 10" tall with the mirror ($550-650) and a matching bed ($575-675). *Furniture from the collection of Gaston and Joan Majeune. Photograph by Renee Majeune.*

Another unique Stevens piece of iron furniture is shown with the bureau. The cabinet measures 8.5" in height ($800-900). The doors and drawers are functional. *Furniture from the collection of Gaston and Joan Majeune. Photograph by Renee Majeune.*

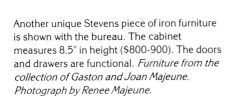

Metal Furniture

The Stevens iron cradle is pictured with a parlor chair. The cradle is 3" long ($375-450). *Furniture from the collection of Gaston and Joan Majeune. Photograph by Renee Majeune.*

Stoves (Metal)

Small kitchen stoves, made of iron, steel, or tin, are among today's most sought after dollhouse collectibles. Because so many different stoves have been made, it is difficult for the collector to distinguish the newer stoves from the genuine antiques. Many small iron cook stoves have been reproduced over the last forty-five years and most were made by using older models as patterns. Some of the names that appear on the fronts of such stoves include Cresent, Queen, Bluebird, and Jewel. The best advice is to be sure to purchase a stove from a reliable dealer.

Most of the early stoves were not really made for dollhouses but were the smallest size of the toy stoves manufactured by a particular company. Other early small stoves were used as salesmen's samples and were produced by the manufacturers of full-sized stoves to promote their products. The first dollhouse sized stoves were modeled after the kitchen stoves, then in use, which used coal or wood for fuel. Some of the companies which produced these stoves were located in Germany but many others were made in the United States. Companies producing these stoves included the Kenton Hardware Co. based in Kenton, Ohio. This company made stoves from 1894 until the middle of this century. Royal and Kent were two of their trade names. Another prolific toy stove manufacturer was the Grey Iron Casting Co. located in Mt. Joy, Pennsylvania. They were in business for over 100 years beginning in 1881. Bluebird and Dot were two of their stove brand names. This company and its successor continued to make toy stoves until 1985. The A.C. Williams Co. of Ravenna, Ohio, also produced iron toys for many years. It may have still been making iron stoves as late as the late 1930s.

As times changed, the cook stove also took on a new look. The gas ranges of the 1920s and early 1930s were made with tall legs which made cleaning under the stove a much easier task than can be accomplished today. Toy stoves followed the new trend and soon most dollhouse stoves were designed in this manner. Hubley Mfg. Co. of Lancaster, Pennsylvania offered a line of iron toy kitchen furniture which included several sizes of gas stoves, ice boxes, tables, and cabinets. Kilgore Mfg. Co. made a much smaller scale (1/2" to one foot) line of iron kitchen pieces for the dollhouse. Their furniture was produced during the 1920s and 1930s (see *American Dollhouses and Furniture From the 20th Century* for more information).

When iron became too expensive to use for dollhouse furnishings, some kitchen pieces were made of sheet metal. An interesting set was manufactured by Marx in a small 1" to one foot scale. The pieces have been found in several colors and appear to date from the late 1930s to the 1940s. Items included in the set were a cabinet, sink, and stove.

More detailed tin kitchen sets were made in Japan circa 1956 which were marketed under the name Linemar. They included a refrigerator, sink, stove, ironer, and washing machine in both 3/4" and 1" to the foot scale. Although most of the firm's metal products were being made in the United States at that time, some toys marketed by the Marx firm were manufactured in Japan.

Another lighter weight kitchen set of metal furniture appears to be made of pressed aluminium. It includes a table, two chairs, stove, refrigerator (4" high), and sink. The pieces are silver with red trim on the two chairs. This set is also unmarked but probably also dates from the 1950s.

With the development of plastic after World War II, most kitchen dollhouse furnishings were made of this new material. During the decade of the 1970s, a new adult interest in dollhouse kits revived the demand for metal kitchen stoves so reproductions of the older models returned to the market place. These continue to be made today.

Shown are a variety of iron toy cook stoves ranging from models of those which used wood to the more recent gas stoves. The Blue Bird stove (top center) is 6" tall and 4" in width. It probably was made by the Grey Iron Casting Co. The Eagle stove was produced by Hubley. Royal was a trade name used by the Kenton Hardware Co. so that stove was probably made by that company (Small wood stoves $50-75, large gas stoves $135-165). *Stoves from the collection of Ray and Gail Carey. Photograph by Gail Carey.*

This iron Kent gas stove measures 4.25" tall. It was made by the Kenton Hardware Co. in Kenton, Ohio ($125-150). *Stove and photograph from the collection of Patty Cooper.*

Metal Furniture 195

Tin cook stoves were also made in sizes small enough to be used in dollhouses. The tin stove on the left and the iron stove on the right are in the 1" to one foot scale while the middle stove would be too large for dollhouse use ($125-$150 depending on accessories). *Stoves from the collection of Ray and Gail Carey. Photograph by Gail Carey.*

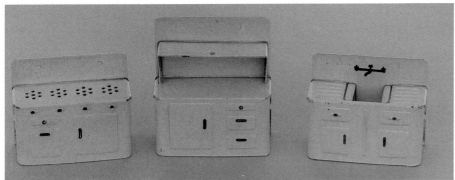

This set of metal kitchen furniture is in a small 1" to one foot scale. The pieces have been found in several colors including gold, white, and ivory. The set was made by Marx circa 1940 ($25-30 each). *Furniture from the collection of Dian Zillner. Photograph by Suzanne Silverthorn.*

This unusual iron kitchen furniture is 3/4" to one foot in scale. The stove is marked "Williams" and may have been made by the A.C. Williams Co. of Ravenna, Ohio. The doors do not function on either piece. Circa 1930s ($65-75 each). *Furniture from the collection of Dian Zillner. Photograph by Suzanne Silverthorn.*

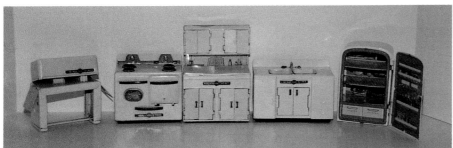

This metal set of kitchen furniture is circa 1956. It was marketed under the name "Linemar." The items are 3/4" to one foot in scale. It was sold by Louis Marx and Co. but made in Japan. The doors are functional ($20-25 each). *Furniture from the collection of Gail and Ray Carey. Photograph by Gail Carey.*

Another set of kitchen appliances also originated in Japan. It is made of enameled tin. The stove is 2" high x 2.75" wide. The products are called "Lucky Stove, Lucky Freezer and Lucky Sink ($20-25 each). *Furniture and photograph from the collection of Marilyn Pittman.*

This unusual kitchen set is made of pressed aluminium. The doors are not functional on this furniture. The refrigerator measures 4" in height ($75-100 set). *Furniture from the collection of Louana Singleton. Photograph by Don Norris.*

Tootsietoy

Tootsietoy dollhouse furniture was produced by the Dowst Brothers Company, located in Chicago, from approximately 1922 to 1937. The furniture was made of a metal alloy in a scale of approximately 1/2" to one foot. Two different sets of the furniture were produced. The sets included furnishings for a living room, bedroom, dining room, kitchen, and bathroom. Cardboard dollhouses were also marketed by Tootsietoy to house the small furniture. See *American Dollhouses and Furniture From the 20th Century* for more information on the Tootsietoy products.

For many years, rumors have circulated among collectors about "French Tootsietoy." Many people use this term, inaccurately, in describing the French metal "Penny Toys," but perhaps there is some connection. The 1925 Tootsietoy catalog shows a "folding go cart" which appears to be identical to the French metal strollers. The example shown is clearly marked "France. S.R." for Simon et Rivollet. The same catalog shows the unusual drop-front desk which, unlike most Tootsietoy pieces, is unmarked and seems to have a strong French "feel." It may be that Tootsietoy imported some pieces from France to accompany their own products.

However, there are other items which seem to more closely fit the term "French Tootsietoy." These pieces are marked "France" and are very similar to specific pieces from the early Tootsietoy line. The furniture is not identical so it was not made from the same molds, but the French pieces correlate to Tootsietoy items in size, color, and style. The French furniture is slightly more delicate and more detailed. It may be that the early Tootsietoy furniture was copied from original pieces made in France.

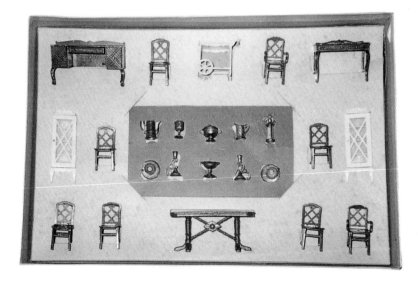

Boxed set of Tootsietoy dining room furniture made from the second design ($250-300). Unlike most of the boxed sets, this one also includes accessories. Each set in this series included different accessories to accompany the furniture. The furniture is 1/2" to one foot in scale. *Furniture and photograph from the collection of Kathy Garner.*

Pictured is one room and its box from the "Add-A-Room" dollhouse sold by Tootsietoy circa 1930s. The cardboard rooms were printed with architectural details on the outside and room decorations on the inside (One room $100-125). *Room from the collection of Ray and Gail Carey. Photograph by Gail Carey.*

Metal Furniture 197

This "Add-A-Room" dollhouse was used to market Tootsietoy dollhouse furniture in the 1930s. Each large box of furniture also included a cardboard room for a dollhouse. The consumer could complete the house by purchasing all the rooms of furniture. There were five rooms in all. *Advertisement from the collection of Louana Singleton. Photograph by Suzanne Silverthorn.*

Metal furniture on the left was made in France while the furniture on the right is from the early Tootsietoy line (circa early 1920s). The similarity in design seems to be more than a coincidence. Both lines of furniture are 1/2" to one foot in scale (French bed $30, buffet $30, Tootsietoy bed $20, buffet $20). *Furniture and photograph from the collection of Patty Cooper.*

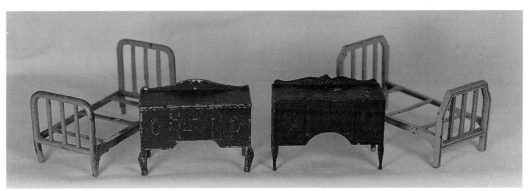

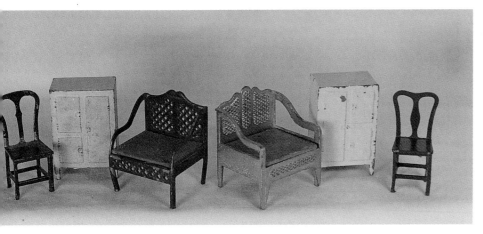

Another example of the "French Tootsietoy" furniture. The pieces on the left are marked "France" while the furniture on the right is from the early Tootsietoy line (French furniture dining room chair $18, ice box $28, and living room chair $30; Tootsietoy dining room chair $12, ice box $18, and living room chair $20). *Furniture and photograph from the collection of Patty Cooper.*

The 1925 Tootsietoy catalog shows a folding go cart which appears to be identical to this one marked "France. S.R." for Simon et Rivollet. Perhaps the American firm imported some items to accompany their regular line of furniture ($45-50). *Go cart and photograph from the collection of Patty Cooper.*

PAPER AND CARDBOARD DOLLHOUSES AND FURNITURE

Dollhouses made of paper or cardboard have been on the commercial market for over a hundred and fifty years. If a person wanted to concentrate only on these types of dollhouses, a varied collection could be assembled.

Most of these houses were made in three different ways. The most often used early design was the book form. Some of these products contained a story and a paper doll family as part of the package. The dollhouse rooms were generally printed flat with furniture and accessories to be cut or punched out to furnish the rooms. Sometimes the pieces were to be glued into place in each room using a number system while many houses were provided with slots to hold the furniture. Other books encouraged young decorators to design their own rooms by placing the furniture to their own liking.

Another method used in the manufacture of paper houses was to design the book so that it opened in a way to display printed rooms that could be used as a background for the paper or cardboard furniture.

The third method used by paper and cardboard house producers was to include the actual pieces of the house in a package, book, or box to be punched or cut-out and assembled. All three of these designs have been used frequently by many different companies both in the United States and abroad.

Usually the furniture for the house was included in the book or package. In order to assemble the three dimensional furniture, glue or tape was sometimes required. More often, the furniture pieces were supposed to stay together with the use of slots and tabs.

Since paper and cardboard houses and furniture could be produced very inexpensively, they provided ideal products to be used in advertising. Although some of the items were nation wide promotions, others were used to advertise local firms and would be hard for a collector to locate. Houses used for premiums usually cost a small amount or could be ordered by mail using proofs of purchase from a product. The furniture was often printed on cards and given away to promote various products.

Many women's magazines included pages of paper dollhouse furniture on their children's pages on an infrequent basis. *Housekeeper Magazine* did a series called "Furniture for the Doll's House" in 1910 and *McCalls Magazine* also published furniture in 1924 called "Sunshine Rooms and Furnishings of Happi House." The same magazine also printed additional furniture in 1926. Newspapers, too, printed paper furniture as an enticement to sell more newspapers. The *Boston Sunday Globe* for September 29, 1895, included an Art Supplement which contained a sofa, arm chair, piano, and piano lamp to be used as dollhouse furniture.

The paper houses that have most often survived the years are the ones that were designed to remain in book form. Because these books were easy to store and the furniture did not have to be taken apart, many of these book houses are still available for collectors today.

Following is a list of some of the paper and cardboard houses and furniture that have been produced during the last one hundred years:

Baby Betty Doll House with Furniture #3926. Made by Whitman Publishing Co. in 1939 (boxed set).

Bradley's Furniture Cut-Outs #8249 designed by Helen M. Fliedner. Milton Bradley Co. Circa 1921. Includes three large sheets of paper furniture to be colored, cut out, and assembled.

Colleen Moore's Doll House Made by Ullman Mfg. Co. #CM3. Castle dollhouse based on the Moore Castle.

Country Playhouse #4099 from Whitman Publishing Co., 1978. Four room house complete with furnishings, paper dolls, and clothing. The company also produced a **Contemporary Playhouse** in 1978 #4098.

Dennison's Crepe & Tissue Paper Doll House Outfit #18. Includes furniture, paper dolls, crepe paper, tissue paper, and patterns for doll dresses. Circa 1900.

Dennison's Four Room Doll House Book. Sold for $1.00 in 1948. Book set up to make a four room house. Furniture and doll cut-outs were in the back of the book.

The Doll's House Model Book No. 244. Published by McLoughlin Bros. in 1905. Included pages to be cut and folded to form four rooms plus the furnishings for the rooms. The rooms were: kitchen, bedroom, nursery, and dining room.

Doll's Open House #220 from Platt and Munk Co., Inc in 1963. House opens to 12" tall. Includes furniture, paper dolls, and clothes.

Dolly Blossom's Bungalow, A Book of Fold-A-Way Toys published by Reilly and Britton, 1917.

Dolly's Furniture made by A. J. Wildman and Son. from Brooklyn, New York. Copyright 1944 and included a kitchen set, cradle set, and bedroom set.

Dolly's Playhouse by McLoughlin Brothers circa 1909. A bungalow house with two rooms on one floor. The front lets down to form a garden. The house is 17" wide, 11" deep, and 7 1/2" tall. (See *American Dollhouses and Furniture From the 20th Century* for more information.)

Fairy tales Cut-Out Dolls and Furniture from Pleasure Books, Inc. Three rooms. Includes King Arthur, Cinderella, and Sleeping Beauty.

Father Tuck's Picture-Building Doll's House Raphael Tuck, 1909 #107.

Fold Away Doll House with punch out furniture. Garden City Books, 1949. Book unfolded to make three rooms. The book came with pieces of furniture to punch out.

Fold-up Furniture made by McLoughlin Bros., Inc. Patented November 5, 1918. Set #4036 included the dining room, bedroom, nursery, library, kitchen, and rustic sets. A total of eight sheets of furniture was in the set. The furniture was to be cut-out and then completed with tabs and notches. See lithographed section.

Folding Doll House made by McLoughlin Brothers circa 1894. The house contained four rooms.

Four Room Doll House Complete With Furniture #7502. Boxed set made by Samuel Lowe Co. in 1943.

The Giant Nine Room Foldaway Doll House with Punch-Out Furniture by Rudolph J. Gutman, 1951.

House For Sale #9042. House contained four rooms on two floors with staircase. Made by Samuel Lowe in 1962. To be punched out.

The House That Glue Built by Clara Andrews Williams with drawings by George Alfred Williams. Published by Frederick A. Stokes, 1905. The book contained seven rooms and furniture and was to be used in book form. The furniture was to be cut out and glued into the proper rooms.

The House That Jack Built published by Stoll & Edwards Co., Inc., New York. Circa 1921. Made of heavy strawboard. Includes garden.

The House That Jack Built #3057. Made by Whitman Publishing Co. in 1933.

The House We Live In published by Gabriel in 1930s. Included accessories and people to be glued into book.

Housekeeping With the Kuddle Kiddies #2140 by Betty Bell Rea. Published by Saalfield Publishing Co., 1936.

Joy-Toys Housekeeping Set circa 1915. Incuded bedroom, dining room, and library furniture to be cut out and folded into place.

Kiddie's Bungalow made by Douglas Mfg. Co. 1922.

Large Size Play House With Auto #154. Art Toy Co. Rochester, N.Y.

Le Pageville Colonial House No. 4. Advertised Le Page Glue. Other houses in this series include a Dutch Colonial, Georgian Colonial, and Cape Cod Cottage. The houses sold for 10 cents each.

Let's Play House #1248 circa 1949. Published by Samuel Lowe Co. Included punch out cardboard furniture and paper dolls.

Mary Francis Housekeeper published by John C. Winston Co., 1914.

McLaughlin XXXX Coffee. Lady paper dolls with pieces of furniture. The coffee company used these items as premiums for their product. The packages included one paper doll, extra garment, hat, and a piece of furniture. The items were made by J. Ottmann Litho Co.

My Doll Family With their Own Home To Cut-Out and Set Up #1985. Whitman Publishing Co., 1955. Book opens into three rooms.

My Dolls' House published by Stecher Litho Co., 1932.

My Dolly's Home published by Arts and General Pub., Lt., London. Circa 1920s.

New Folding Doll House made by McLoughlin Brothers in 1909. The front could be let down to show the inside of the house. The house was 19 1/2" tall. The house had two rooms on two floors.

Our New Home published by Gabriel in 1930. #890.

Shirley Temple and Her Playhouse #1739 by Saalfield Publishing Co., 1935.

Two Story Doll House #2145. Published by Lowe in the 1940s.

Universal Toy and Novelty Mfg. Co. produced paper furniture circa 1900 that included furniture for a dining room, kitchen, parlor, and bedroom. The pieces were to be cut out and assembled.

Whitman Publishing Co. made a series of sets of furniture in the early 1930s. Included are: #968A Kitchen, living room, and bath with complete furnishings, 1932. **Let's Make Furniture**: #995 Dining room, bedroom, and kitchen, 1934. Living room, bath, and parlor #995 also in 1934. **Peek-A-Boo Playhouse** #979, 1933.

The Wonderland Doll's House by Schmidt Lithograph Company, San Francisco, California. Made of cardboard, opens to form 3-D room settings. Includes furnishings.

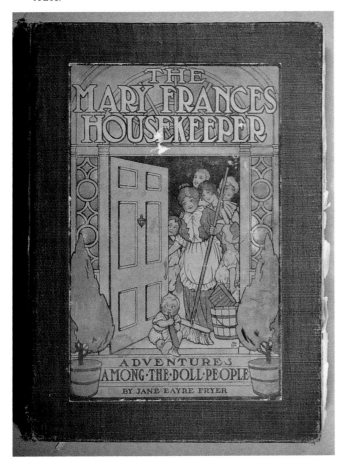

The Mary Frances Housekeeper book was published by the John C. Winston Co. in 1914. This "house" book included very nice paper dolls as well as three dimensional furniture ($75-100). *Book from the collection of Elaine Price. Photograph by Suzanne Silverthorn.*

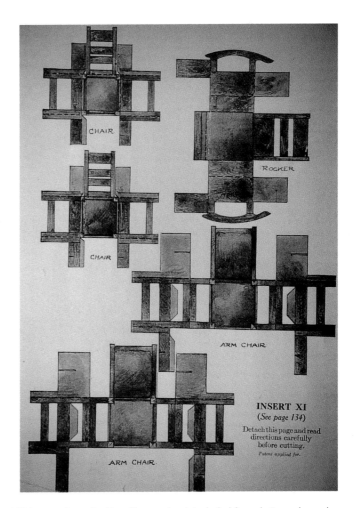

This page from the Mary Frances book included four chairs and a rocker to be assembled. *From the collection of Elaine Price. Photograph by Suzanne Silverthorn.*

200 Paper and Cardboard Dollhouses and Furniture

Fold Away Doll House with Punch Out Furniture published by Garden City Books, 1949. Designed by Catherine Barnes. The book is heavy cardboard and it unfolds to make three rooms with walls and floors. The furniture was to be punched out and assembled ($40-45). *Book and photograph from the collection of Ruth Petros.*

The house and furniture were made of a light cardboard covered with crepe paper. The paper dolls were similar to others sold by Dennison that were to be dressed with crepe paper. Circa 1900 ($75-100). 13" tall x 12" wide x 12" deep. *House and photograph from the collection of Sharon Unger.*

Box containing Dennison's Crepe & Tissue Paper Doll House Outfit #18. The contents included one "substantial" crepe paper house with windows, blinds, doors, and chimney; set of furniture consisting of bed, chairs, table, bureau, picture, rugs; two jointed dolls, four small rolls of crepe paper, tissue paper, gold and silver paper, lace, stars, and three patterns for doll dresses. *Box and photograph from the collection of Sharon Unger.*

The Wonderland Doll's House made by the Schmidt Lithograph Company, San Francisco. The advertising stated it was a real doll mansion with twelve complete rooms and lithographed in ten colors. The house was made of heavy cardboard and could be folded flat. It was sold by Macys in 1931 ($50-75). *House and photograph from the collection of Sharon Unger.*

Paper and Cardboard Dollhouses and Furniture 201

The flat Wonderland house opened in two layers in the back to form 3-D room settings. *House and photograph from the collection of Sharon Unger.*

Boxed *Shirley Temple and her Playhouse* #1739 from Saalfield Publishing Co., 1935 ($175-200). *From the collection of Elaine Price. Photograph by Suzanne Silverthorn.*

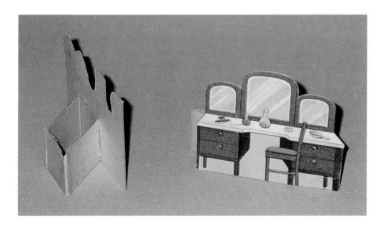

The cardboard furniture for the Wonderland house was one dimensional with a cardboard standard on the back. *Furniture and photograph from the collection of Sharon Unger.*

The Shirley Temple house and furniture were to be assembled after purchase. The contents included a Shirley Temple paper doll, clothing, and 3-D furniture. *From the collection of Elaine Price. Photograph by Suzanne Silverthorn.*

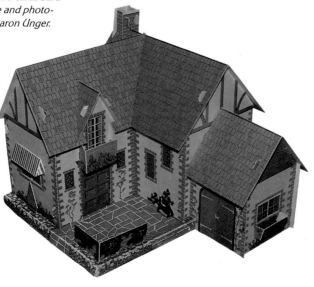

This Built-Rite cardboard house is circa late 1930s ($60-80). The company which made the house began as the Warren Paper Products Co. in Lafayette, Indiana, in the early 1920s. They began the manufacture of toys during the mid-1930s. *House from the collection of Nanci Moore. Photograph by Roy Specht.*

202 Paper and Cardboard Dollhouses and Furniture

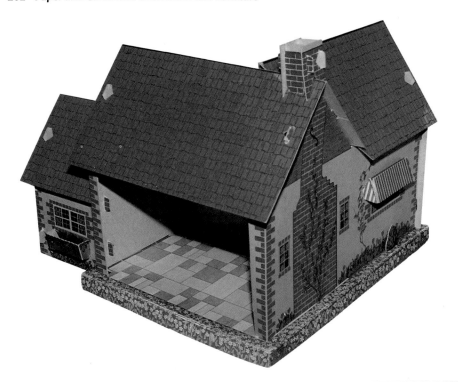

The Built-Rite house measures 11" high x 13" wide x 10" deep. *House from the collection of Nanci Moore. Photograph by Roy Specht.*

Built-Rite also produced cardboard furniture for its houses. This was the box made for the early dining room furniture circa mid-1930s. *Box and photograph from the collection of Patty Cooper.*

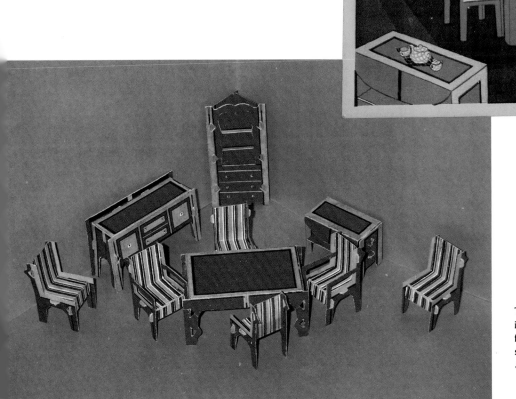

This cardboard Built-Rite dining room furniture is part of the early set ($4-6 each). The furniture is approximately 3/4" to one foot in scale. *Furniture from the collection of Louana Singleton. Photograph by Don Norris.*

Paper and Cardboard Dollhouses and Furniture 203

This early living room set is circa mid-1930s ($4-6 each). *Furniture from the collection of Louana Singleton. Photograph by Don Norris.*

The bedroom pieces, like all the other Built-Rite furniture, was made to be punched out and assembled by the consumer ($4-6 each). *Furniture from the collection of Louana Singleton. Photograph by Don Norris.*

The early Built-Rite kitchen furniture was brightly colored as are all the older pieces ($4-6 each). *Furniture from the collection of Louana Singleton. Photograph by Don Norris.*

Built-Rite cardboard furniture was featured in this Sears-Roebuck catalog in the mid 1930s. The bathroom pieces included a bathtub, toilet, sink, and hamper. The house was probably also made by Built-Rite. *Catalog from the collection of Linda Boltrek. Photograph by Suzanne Silverthorn.*

204 Paper and Cardboard Dollhouses and Furniture

Cardboard "New Modern Home Seven Rooms Doll House" made by O.B. Andrews Co. in Chattanooga, Tennessee, circa 1935 (House and furniture $125-150). 16" tall x 30" wide x 13" deep. *House and photograph from the collection of Patty Cooper.*

The house included seven rooms complete with a fireplace and stairway. Printed floor coverings also added interest to the house. Sets of furniture were supplied for the kitchen, dining room, parlor, bathroom, sun parlor, and two bedrooms. *House and photograph from the collection of Patty Cooper.*

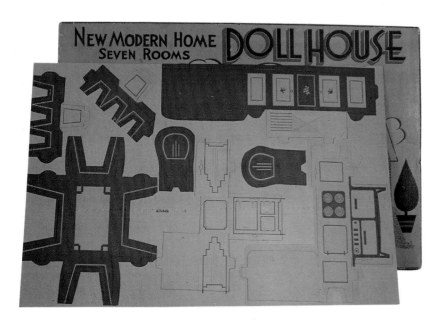

The cardboard furniture was made to be assembled by the consumer. Kitchen pieces included a range, table, chairs, and cabinet. *Furniture and photograph from the collection of Patty Cooper.*

Paper and Cardboard Dollhouses and Furniture 205

Furniture for two bedrooms was packaged with the house but both sets were alike except for the coloring. *Furniture and photograph from the collection of Patty Cooper.*

Living room furniture was designed to include an overstuffed sofa and chairs. A radio was also a part of the living room furnishings. *Furniture and photograph from the collection of Patty Cooper.*

Box for "Four Room Doll House Complete With Furniture" manufactured by the Samuel Lowe Co. in Kenosha, Wisconsin, in 1943 (Boxed house and furniture $75-95). *Box and photograph from the collection of Betty Nichols.*

206 Paper and Cardboard Dollhouses and Furniture

The Lowe cardboard four-room house (#7502) measures 13" high (not including chimney) x 14.75" wide x 8.25" deep. *House and photograph from the collection of Betty Nichols.*

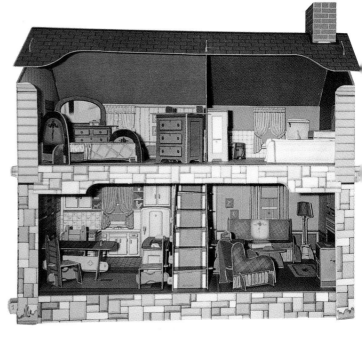

Both the house and the cardboard furniture had to be assembled by the consumer. *House and photograph from the collection of Betty Nichols.*

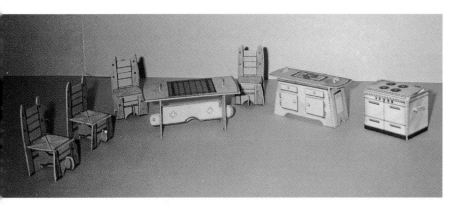

The cardboard Lowe kitchen pieces included a table, four chairs, buffet, and a stove. The refrigerator and sink were both printed on the kitchen walls. *Furniture and photograph from the collection of Betty Nichols.*

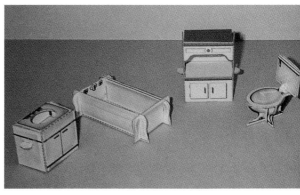

The Lowe bathroom was furnished with a sink, bathtub, toilet, and cabinet. The furniture seems to be too large for the house. *Furniture and photograph from the collection of Betty Nichols.*

Most of the same furniture was also packaged by Lowe in a box which did not include a house. It was labeled "Doll House Furniture/No Cutting No Pasting No Tools Needed/Made of Heavy Sturdy Built Board #2574." (Boxed set $40-60). *Box from the collection of Mary Lu Trowbridge Photograph by Bob Trowbridge.*

Paper and Cardboard Dollhouses and Furniture 207

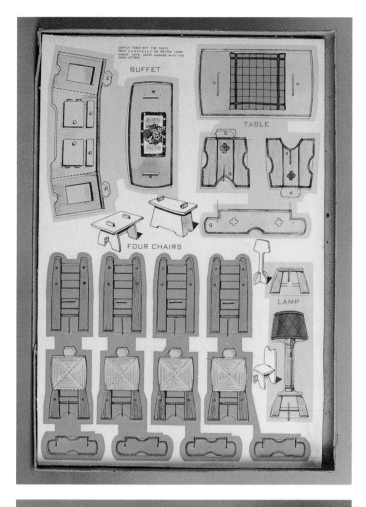

This set of Lowe cardboard furniture included pieces to furnish three rooms. The earlier kitchen pieces were labeled as dining room furniture and the kitchen stove was eliminated. *Furniture from the collection of Mary Lu Trowbridge. Photograph by Bob Trowbridge.*

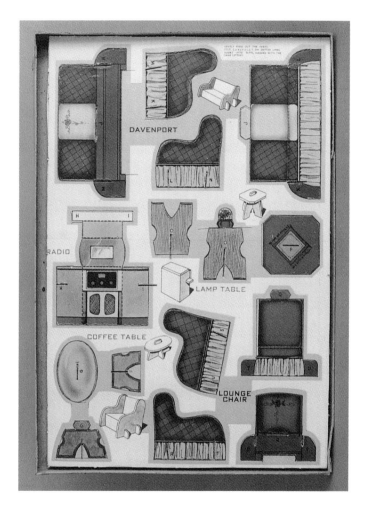

The Lowe living room included a radio as well as a davenport, lounge chair, lamp table, and coffee table. *Furniture from the collection of Mary Lu Trowbridge. Photograph by Bob Trowbridge.*

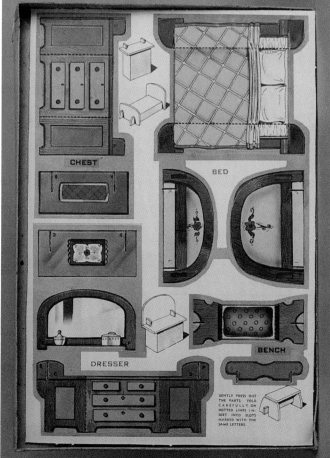

Uncut bedroom furniture made by Lowe. *Furniture from the collection of Mary Lu Trowbridge. Photograph by Bob Trowbridge.*

208 Paper and Cardboard Dollhouses and Furniture

Trixy four-room portable dollhouse measuring 13" high x 17" wide. The cardboard house was made by the Durrel Company in Boston, Massachusetts, circa 1920s. The inside of the house is quite plain except for curtains printed around the windows ($65-75). *House and photograph from the collection of Linda Boltrek.*

This cardboard house was made by the Gable House and Carton Co. located in New York City. It was called the "Gable Villa Doll House" according to the company's advertising in 1915 ($65-75). *House from the collection of Joanie Searles. Photograph by Ernie Hanni.*

The back opening Gable house originally sold for only twenty-five cents and was quite small being approximately 12" wide. *House from the collection of Joanie Searles. Photograph by Ernie Hanni.*

Box and cardboard house called "The House That Jack Built." It was made by Stoll & Edward Co., Inc., New York, in 1921. The house contained only one room ($50-75). *House and photograph from the collection of Sharon Unger.*

Paper and Cardboard Dollhouses and Furniture 209

The front of the house is adjustable and can be removed. There is also a garden to accompany the house which is made of heavy strawboard. 16" tall x 15" wide x 11" deep. *House and photograph from the collection of Sharon Unger.*

The Play-Town Doll House No. 200 was manufactured by Sutherland Paper Company in Kalamazoo, Michigan ($125-135). 18" high x 23" wide x 12.5" deep. *House from the collection of Nanci Moore. Photograph by Roy Specht.*

The inside of the Sutherland house contains five rooms with printed floor coverings. The front is removable. Circa 1930. *House from the collection of Nanci Moore. Photograph by Roy Specht.*

Concord cardboard dollhouse circa 1940. The company was located in New York City ($75-100). 15.5" high x 29.5" wide x 10.5" deep. *House from the collection of Nanci Moore. Photograph by Roy Specht.*

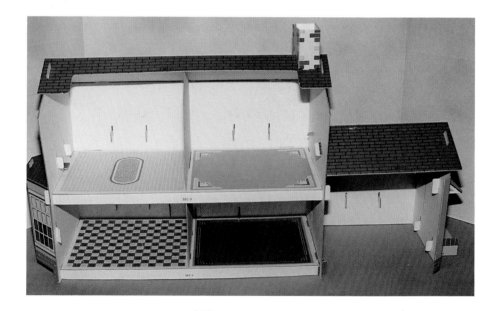

The Concord house contained five rooms with printed floor coverings. *House from the collection of Nanci Moore. Photograph by Roy Specht.*

Sears Happi-Time dollhouse advertised in their 1946 Christmas catalog. The cardboard house was advertised as being made of fiberboard, the heaviest the company had ever used for this type of house ($85-110). 19.5" high x 28.5" wide x 18" deep. *House from the collection of Nanci Moore. Photograph by Roy Specht.*

Paper and Cardboard Dollhouses and Furniture 211

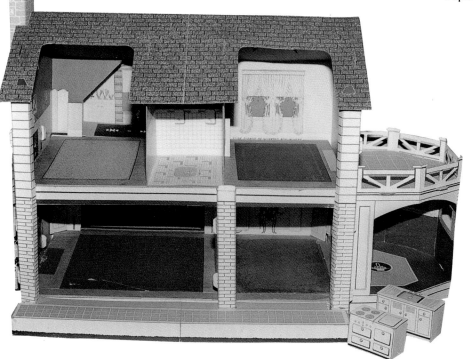

The Happi-Time house included six rooms furnished with Renwal furniture except for the kitchen which came with cardboard stove, sink, and refrigerator (not pictured). A similar house was sold by Sears in 1949 which included a garage. *House from the collection of Nanci Moore. Photograph by Roy Specht.*

This corrugated cardboard house was manufactured by the Atlantic Container Corp. from Long Island City, New York. The house was labeled a Fibre-Bilt Toy, "Sub-bur-bannette," Doll House #106 ($50-60). *House from the collection of Becky Norris. Photograph by Don Norris.*

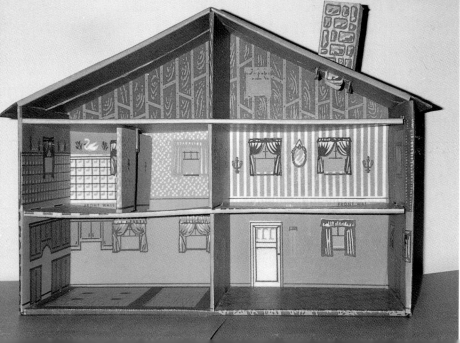

The house contains five rooms plus two attic rooms. The inside includes printed walls and floors. 16" tall (to roof peak) x 21.5" wide x 9" deep. *House from the collection of Becky Norris. Photograph by Don Norris.*

212 Paper and Cardboard Dollhouses and Furniture

Instant Play House (#0504). made by the Winthrop-Atkins Co., Inc. in Middleboro, Massachusetts. The heavy cardboard house came with a sink/stove, refrigerator, sofa, chair, and television also made of cardboard. The house was to be used with the small 4" dolls of the time including Liddle Kiddles (Mattel, Inc). Circa mid 1960s ($25-35). 7.25" to the roof peak x 11.5" wide x 7.75" deep. *House from the collection of Becky Norris. Photograph by Don Norris.*

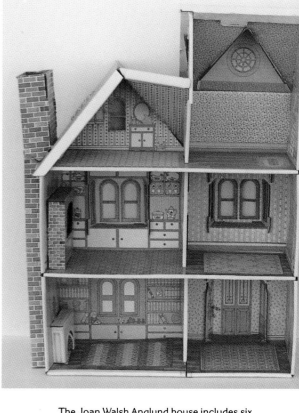

This is a Joan Walsh Anglund Doll House made in 1979. Joan Walsh Anglund dolls were to be used with the house ($25-30). 23" high x 19.5" wide x 9.5" deep. *House from the collection of Dian Zillner. Photograph by Suzanne Silverthorn.*

The Joan Walsh Anglund house includes six rooms plus an attic. The rooms are brightly decorated with wall and floor designs. *House from the collection of Dian Zillner. Photograph by Suzanne Silverthorn.*

Advertising dollhouses are especially appealing to collectors. This cardboard Cottage was a premium used by Cloverbloom Butter circa 1930 ($65-75). 12" high x 12" wide x 10" deep. *House and photograph from the collection of Ruth Petros.*

Paper and Cardboard Dollhouses and Furniture 213

The furniture for the Cloverbloom Cottage was printed on the inside of the butter packages ($6-8 each). It was to be cut out and assembled by the consumer after the butter was removed from the package. *Furniture and photograph from the collection of Ruth Petros.*

Another advertising dollhouse is this cardboard "Miss Sunbeam's Doll House." The house is missing the awnings and chimney. Circa late 1940s. Quality Bakers of America own the "Sunbeam" bread trademark ($65-75). 11" tall x 36" wide x 12" deep. *House from the collection of Becky Norris. Photograph by Don Norris.*

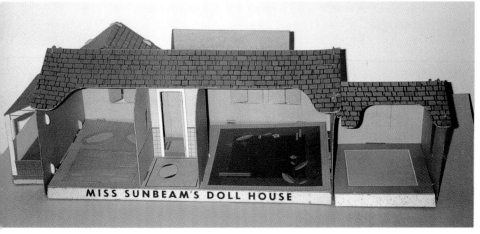

The Sunbeam house includes four rooms which feature unusual printed floor coverings. *House from the collection of Becky Norris. Photograph by Don Norris.*

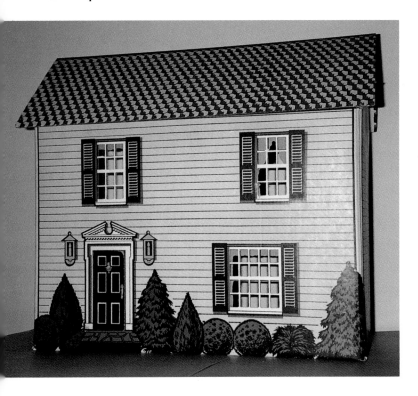

Another more recent advertising dollhouse is this one used as a premium by Kool Aid circa 1960s ($35-45). 22.5" high x 24" wide x 12.25" deep. *House from the collection of Becky Norris. Photograph by Don Norris.*

The inside of the Kool Aid house contains four rooms with very colorful lithography on the walls. The house is labeled on the back roof, "The Smiling Pitcher is a Registered/Trademark of General Foods Corp." *House from the collection of Becky Norris. Photograph by Don Norris.*

Armstrong Cork Company, Floor Division, in Lancaster, Pennsylvania, issued several cardboard room settings with a few pieces of cardboard furniture circa 1940s. The rooms featured copies of Armstrong floor coverings to promote the products ($20-25 each room). Rooms measure 4.25" tall x 8.75" wide x 8.75" deep. *Dining room from the collection of Becky Norris. Photograph by Don Norris.*

Cardboard dollhouse furniture was also used for advertising purposes through the years. This living room set is 1/2" to one foot in scale and was used to promote Shotwell's-Chicago. On the back of each piece of furniture is the statement that "There are six different pieces of living room furniture in this set." The matching chairs are unmarked and may not be part of the original set although they are made in the same colors and material as the other items ($20-25 set). *Furniture from the collection of Dian Zillner. Photograph by Suzanne Silverthorn.*

Paper and Cardboard Dollhouses and Furniture 215

Joy-toys Housekeeping Set produced by Oxford-Print, Boston, Massachusetts, circa 1915 ($100-150 for set of three rooms). This Joy-Toys package under the name "Dolly Furniture" was also used for a premium for anyone subscribing to *Little Folks* magazine in 1918. *Set from the collection of Becky Norris. Photograph by Don Norris.*

The Joy-toys room size rug measures 10.5" x 13.25" and the buffet is 3.25" tall. The furniture is made of a light weight cardboard. *Dining room from the collection of Becky Norris. Photograph by Don Norris.*

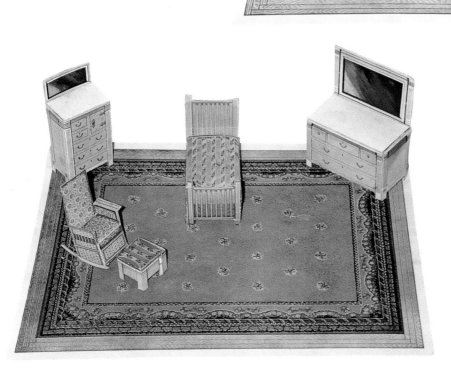

The Joy-toys bedroom includes five pieces of furniture along with a rug. *From the collection of Becky Norris. Photograph by Don Norris.*

216 Paper and Cardboard Dollhouses and Furniture

The Joy-toys library furniture could also be used to furnish a dollhouse living room. All of this furniture would fit nicely in a Dunham's Cocoanut house although the furniture is not of the same period. *From the collection of Becky Norris. Photograph by Don Norris.*

These pieces of thin cardboard furniture were made by the Worth While Mfg. Co. circa 1915 ($8-10 each). The trade name for the furniture was Toyville Furniture. This set #201 originally came with eleven pieces of furniture. Besides the items illustrated, the set included a grandfather clock, bed, four chairs, and a library table. The sofa is 3" wide and 2.25" tall. *Furniture from the collection of Becky Norris. Photograph by Don Norris.*

PLASTIC FURNITURE AND METAL HOUSES

RELIABLE

When plastic dollhouse furniture became popular in the United States in the late 1940s and 1950s, it also gained a following in Canada through the production of furniture by the Reliable Plastics Co, Ltd. This firm was based in Toronto, Canada. The 3/4" to one foot scale furniture was sold either by the piece or in room sets. The early boxes could be set up as rooms similar to the Renwal rooms made in the United States. Furniture was made to furnish the living room, bedroom, dining room, nursery, bathroom, kitchen, and laundry area. Several other school and playground items were also produced. Each piece of furniture was marked "Reliable." The Reliable furniture was made from the late 1940s into the 1950s and is now quite in demand by plastic collectors everywhere.

Reliable was licensed by the Ideal Toy Co. (based in New York) to copy some of its products. Several of the Ideal doll molds were used by Reliable. The same agreement must have applied to the plastic dollhouse pieces as many of the items marked with the Reliable name have come from the Ideal dollhouse furniture molds. These included kitchen and bathroom pieces plus the floor radio, grand piano, and dining room chairs (like the Ideal kitchen chairs).

Furniture made by Reliable included the following pieces: Living room: Chesterfield sofa, floor radio, chair, end table, coffee table, fireplace, floor lamp, table lamp, and piano with bench. Dining room: Duncan Phyfe type table, four chairs, buffet, china cabinet, and tea cart. Bathroom: wash basin, toilet, bathtub, and hamper. Nursery: highchair, cradle, doll carriage, larger doll carriage, bathinette, playpen, stroller, potty chair, rocking chair, and small plastic doll 2 1/2" tall with movable arms and legs. Kitchen: refrigerator, sink, stove, table, and four chairs. The appliances feature the "Reliable" name on the fronts of each piece. Bedroom: chest of drawers, night table, dresser with mirror, vanity with mirror, stool, bed, and rocker. Laundry room: sewing machine and wringer washer. Other Reliable products include a teacher's desk, swivel desk chair, student desks, tricycle, merry-go-round, swing, lawn mower, carpet sweeper, and bunk beds with ladder. Several different baby dolls were made for the Reliable furniture. These included the hard plastic 2 1/2" tall model with movable arms and legs, two different sized babies made of a rubber-like material, and another molded plastic doll that was made in both pink and black.

The early furniture from Reliable was packaged in room settings similar to Renwal. Pictured is the outside of the living room. The room measures 6.5" high x 9.5" wide x 9.5" deep. *Room and photograph from the collection of Mary Harris.*

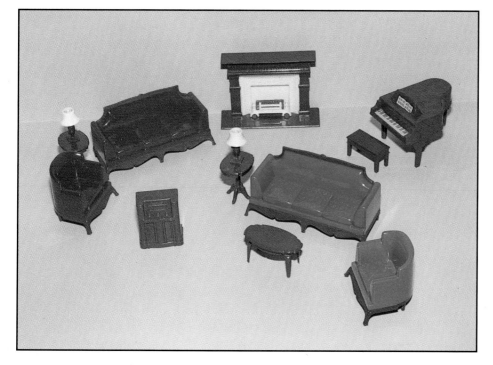

The Reliable 3/4" to the foot scale living room furniture included a Chesterfield sofa ($25-30), end table ($20-22), coffee table ($15-20), chair ($20-25), fireplace ($34-45), radio ($15-20), table lamps ($20-25), piano and bench ($35-45), and a floor lamp which is not pictured. *Furniture from the collections of Nanci Moore and Roy Specht. Photograph by Roy Specht.*

The inside of the cardboard living room was printed with draperies, a rug, and wall decorations to give a realistic background setting for the furniture. ($175-225). *Furniture and photograph from the collection of Mary Harris.*

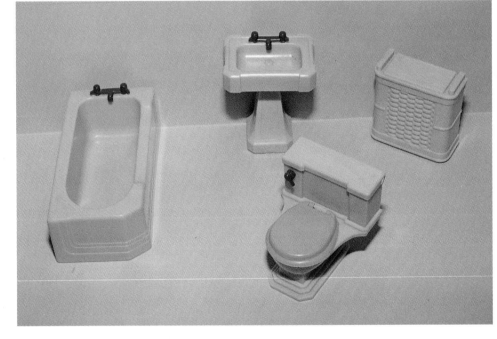

Bathroom plastic furniture by Reliable included four pieces: bathtub ($15-25), wash basin ($15-20), toilet ($15-20), and hamper ($15-20). *Furniture from the collections of Nanci Moore and Roy Specht. Photograph by Roy Specht.*

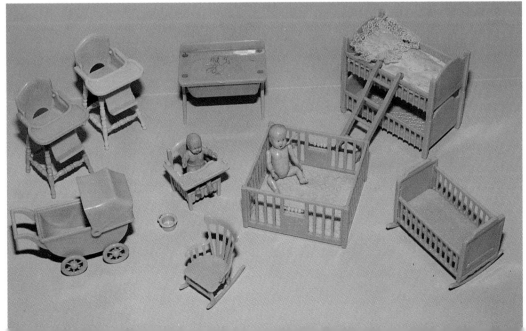

Reliable furniture for the nursery included the following pieces: high chair ($25-35), two sizes of doll carriages ($40-45), crib ($45-55), rocking chair ($15-20), potty chair ($30-35), playpen ($25-35), bathinet ($30-40), bunk beds ($75-100), and baby dolls ($35-45). Only one doll carriage is pictured. *Furniture from the collections of Nanci Moore and Roy Specht. Photograph by Roy Specht.*

Plastic Furniture and Metal Houses 219

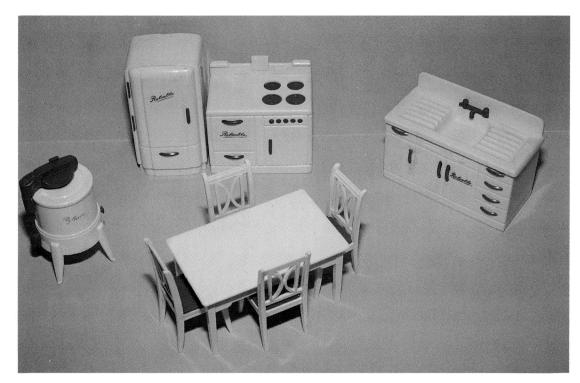

Reliable kitchen furniture included a refrigerator ($25-30), stove ($25-30), sink ($25-30), table and four chairs ($45-55). The three appliances have the "Reliable" name on their front panels. The washing machine ($35-45) was featured in the Reliable advertising as a "Gay Monday" piece along with a sewing machine. *Furniture from the collections of Nanci Moore and Roy Specht. Photograph by Roy Specht.*

Pictured is one of the early Reliable boxes along with kitchen pieces (Boxed set $175-225). *Furniture and photograph from the collection of Mary Harris.*

The dining room furniture included a Duncan Phyfe type table ($25-35), buffet ($35-45), china cabinet ($35-45), four chairs ($10-15), and a serving cart ($45-55). Also pictured is the sewing machine ($35-40). *Furniture from the collections of Nanci Moore and Roy Specht. Photograph by Roy Specht.*

220 Plastic Furniture and Metal Houses

A later box which holds the dining room pieces did not include a room setting (Boxed set $175-200). *Furniture and photograph from the collection of Roy Specht.*

Other Reliable items include a lawn mower ($35-45), tricycle ($35-45), swing ($50-60), wagon ($45-55), and a watering can ($10-15). *Pieces and photograph from the collection of Geraldine Raymond-Scott.*

A variety of other pieces were produced by Reliable. Included were a teacher's desk ($35-40), swivel chair ($20-25), student desks ($25-30), and a merry-go-round ($65-75). *Furniture and photograph from the collection of Roy Specht.*

The Reliable bedroom furniture included bed ($35-45), vanity ($25-35), stool ($10-15), chest of drawers ($25-35), dresser ($25-35), night table ($10-15), and rocker ($15-20). *Furniture from the collections of Nanci Moore and Roy Specht. Photograph by Roy Specht.*

Plastic Furniture and Metal Houses   221

MISCELLANEOUS FURNITURE AND HOUSES

Ardee Plastics, Inc.

In 1946 Ardee Plastics, Inc. of New York became an early manufacturer of plastic dollhouse furniture in the 3/4" to one foot scale. Their furniture was made of Lumarith, a product of the Celanese Plastics Corporation. It is known that the firm produced bedroom and living room pieces as well as a table and four chairs. The furniture has tended to warp, as the years have passed, so much of it has probably been destroyed.

Living room items included a sofa, chair, bookcase, fireplace, floor lamp, table lamp, and coffee table. The bedroom furniture included a bed, chest of drawers, night stand, vanity with mirror, and lamp.

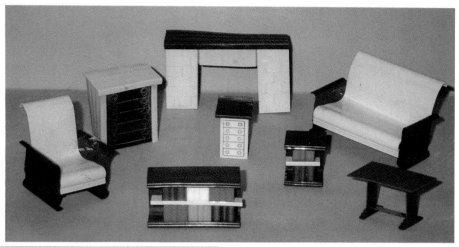

Ardee Plastics, Inc. living room furniture from 1946 in 3/4" to one foot scale. Pieces included a sofa ($10-15), chair ($10-12), bookcase ($5-10), fireplace ($5-10), coffee table ($5-10), lamp table ($5-10), and lamps ($5-10). *Furniture and photograph from the collection of Mary Brett.*

Ardee bedroom furniture included a bed ($15-20), vanity ($10-15), floor lamp ($5-10), chest of drawers ($10-15), and nightstand ($8-10). *Furniture from the collection of Karen Evans. Photograph by Don Norris.*

Auburn

Auburn Rubber was founded in Auburn, Indiana, in 1913 as the Double Fabric Tire Corporation. The company made their first toys in 1935. The toy division closed in 1968 after a move to Deming, New Mexico, in 1960. The firm marketed a miniature Auburn Dollhouse during the 1950s. The dollhouse came in small room units. Each unit could be connected to the others in order to make a complete house. The rooms were made of plastic and cardboard and the original boxes were marked "Auburn Safe Play Toys" in a circle. Rooms included a living room (sofa, two chairs, television, and table), kitchen-dinette (table, two chairs, refrigerator, sink, and stove), utility room-playroom (ping pong table, television, two chairs, washer, dryer, and table), and bedroom-bath (bed, vanity, dresser, two chairs, bathtub, toilet, and large room divider sink). Each room measured approximately 5" tall x 7" wide x 4.25" deep.

Three units of the Auburn Dollhouse circa 1950s ($45-60 each). 5" high x 7" wide x 4.25" deep. *Houses from the collection of Nanci Moore. Photograph by Roy Specht.*

The Auburn rooms were made of plastic and cardboard and the furniture was plastic. Pictured are the utility room/playroom; bedroom/bath; and the living room. *Furniture from the collection of Nanci Moore. Photograph by Roy Specht.*

Blue Box

Blue Box plastic dollhouse furniture was popular during the late 1960s and early 1970s. It was made in Hong Kong and was featured in mail order catalogs issued by Sears, J. C. Penney, and Montgomery Ward during this period. The furniture was mostly in the small scale of 1/2" to one foot although a few pieces were larger. Some of the furniture was made from Jennys Home designs produced by Spot-On-Models Limited, a Triang Product, from Great Britain.

From 1968 to 1971, sets of furniture were sold with various vinyl cases which were to be used as houses for the furniture. The prices for these items were very reasonable. A more elaborate Blue Box set was marketed in 1970. The house contained five rooms and was furnished with both 1/2" and 3/4" to the foot scaled furniture in the same house. Outside furniture, as well as a fence and car, were also included in the set.

Although the Blue Box furniture was made very cheaply, the doors and drawers on the pieces functioned and youngsters could experience the pleasure of owning a furnished dollhouse for a very reasonable price.

Vinyl "suitcase" style dollhouse containing two rooms. The house, marked "© Ideal Toy Corp.," was furnished with Blue Box furniture and is circa early 1970s. A similar house was sold in the Montgomery Ward Christmas catalog in 1970 as an "English Cottage." ($35-45 furnished). 9" high x 12" wide x 8.5" deep. *House and photograph from the collection of Patty Cooper.*

The inside of the "Blue Box" dollhouse included plastic living room, dining room, and bedroom furniture. *House and photograph from the collection of Patty Cooper.*

Plastic Furniture and Metal Houses 223

This Blue Box advertisement pictures a plastic house in the 1/2" to one foot scale furnished with Blue Box furniture. The layout as shown measures 21" x 15" and is circa 1970. *From the collection of Linda Boltrek. Photograph by Suzanne Silverthorn.*

Boxed set of Blue Box plastic furniture circa 1970. Includes pieces for a bedroom, bathroom, nursery, living room, kitchen, and dining room. The box is labeled, "Blue Box Doll's House Furniture. Made in Hong Kong." The furniture ranges in size from a large 1/2" to a small 3/4" to one foot in scale ($30-35). *Furniture from the collection of Dian Zillner. Photograph by Suzanne Silverthorn.*

Casablanca Products, Incorporated

Casablanca Products, Incorporated was located in Bridgeport, Connecticut. The company was first listed in the Bridgeport Directory in 1947 with Sam Levin as president. The company name changed in 1953 to Connecticut Plastics with Irving Rosenzweig as president. Although it was still in business in Bridgeport in 1954, there is no record of the company in that city after that time.

Of interest to dollhouse collectors is the plastic dollhouse furniture made by the firm circa 1947. Although only the bedroom set is pictured here, it is probable that other rooms of furniture were made.

Casablanca Products, Incorporated, located in Bridgeport, Connecticut, manufactured several sets of plastic dollhouse furniture in the late 1940s. Pictured is the box which housed the bedroom pieces ($85-100 boxed set). *Furniture from the collection of Louana Singleton. Photograph by Don Norris.*

Casablanca furniture was 3/4" to one foot in scale. The bedroom furniture consisting of bed, chest of drawers, nightstand, vanity, and bench is shown. *Furniture from the collection of Louana Singleton. Photograph by Don Norris.*

Deluxe Reading Corporation

The Deluxe Reading Corporation, located in Elizabeth, New Jersey, was an active participant in toy production during the late 1950s and 1960s. Their products were sold under several different names including Deluxe Topper, Topper Corp., Topper Toys, Deluxe Toy Creations, and Deluxe Premium Corporation. Many of their toys were sold through super markets or were used for premiums. One of their most popular dolls was the Penny Brite 8" vinyl doll issued in 1964.

In 1963 the firm marketed a dollhouse called "Debbie's Dream House" made of cardboard and plastic. The house was a very modern, one-story ranch style model with play access through the open top. It contained an open-plan living room, dining room, and kitchen, with a hallway which led to a bathroom and two bedrooms. One of the bedrooms had a closet with sliding doors. The furniture was a very modern sixties style, 3/4" to one foot, and was made of plastic with some cardboard parts. The flooring was printed on the base, and pictures were printed on the walls. One interesting characteristic of this set is that everything was supplied in threes instead of the more usual fours. There were three living room chairs, three chairs to the dining table, and three bar stools in the kitchen.

The Debbie house required batteries which allowed the fireplace and lamp to "glow." The doorbell was also operational. The house came unassembled. The base of the assembled house was 36" wide x 21" deep.

"Debbie's Dream House" was made by the Deluxe Reading Corporation and came unassembled. The house was made of plastic and cardboard and included a fireplace and lamp made to glow with batteries. Plastic furniture was included with the house. The box reads "Debbie's Dream House t.m. Another Fine Toy by Deluxe. Deluxe Reading Corporation. Elizabeth, N.J." Circa 1963 ($45-65). Base: 36" wide x 21" deep. *House and photograph from the collection of Patty Cooper.*

Eagle Toy Company

The Eagle Toy Company, located in Montreal, Canada, made many different styles of dollhouses during the 1950s. The firm's two-story metal houses resembled the Jayline houses produced in the United States during this same time period. These houses were quite small and would have been furnished with 1/2" to one foot furniture. The company's unusual hexagonal house was in a scale of 3/4" to one foot and was made of both fiberboard and metal.

The Eagle Toy Company was located in Montreal, Canada, and the firm produced several different models of dollhouses through the 1950s. This metal house is 1/2" to one foot in scale and was made in 1959 ($75-85). 13.5" high x 18.5" wide x 7.5" deep. *House and photograph from the collection of Mary Harris.*

Plastic Furniture and Metal Houses 225

The inside of the metal house made by the Eagle Toy Co. contained five rooms. *House and photograph from the collection of Mary Harris.*

Pictured is another small metal house produced by the Eagle Toy Co ($75-85). 15.5" to chimney top x 18.5" wide x 7" deep. *House and photograph from the collection of George Mundorf.*

The inside of this Eagle house features a different decor in its five rooms. *House and photograph from the collection of George Mundorf.*

Besides metal houses, Eagle Toy Co. also produced models made of fiberboard and tin. Pictured is a hexagon-shaped house on casters so it can turn easily for access to its six rooms ($75-100). 7.5" high x 21" wide. *House and photograph from the collection of Mary Harris.*

Ideal

The Ideal plastic dollhouse furniture came on the market in 1947. It was produced by the Ideal Novelty and Toy Co. located in Hollis, New York. The furniture is considered to be 3/4" to one foot in scale but some of the pieces are slightly larger. The furniture was sold in room sets which included the living room, dining room, bathroom, bedroom, kitchen, nursery, and patio. Extra pieces of furniture were also made and many of these items could be purchased at the local "dime stores." Ideal also produced baby dolls and a family of dolls to accompany the plastic furniture.

Another line of Ideal dollhouse furniture was manufactured in 1950. This larger set of plastic furniture was called Young Decorator and was scaled approximately 1 1/2" to one foot. This line of furniture included pieces for a living room, dining room, bedroom, bathroom, and nursery as well as several extra items.

The most unusual Ideal plastic dollhouse furniture is the Petite Princess line marketed in 1964. Furniture was included for a bedroom, living room, dining room, and extra pieces that could be used in a music room. This "fantasy" furniture along with the Princess Patti line (which included bathroom, kitchen, and extra items) introduced in 1965 are in constant demand by today's collectors. See *American Dollhouses and Furniture From the 20th Century* for more information on the Ideal products.

Vinyl "suitcase" type house made by Ideal. The house dates from 1969. The plastic furniture included with the house is marked "1969 Ideal Toy Corp. Hong Kong." (Furnished $100-125). 17.5" high x 20" wide. *House and photograph from the collection of Roy Specht.*

Plastic Furniture and Metal Houses 227

The inside of the Ideal two story house includes four separate areas for furniture plus a front which drops down to become a patio. *House and photograph from the collection of Roy Specht.*

The Ideal vinylite family dolls were produced only in 1949. The parents originally retailed for 29 cents and were 5" tall. The children measured 3" high and were priced at 19 cents. The dolls have movable arms and painted clothes. The dolls are quite scarce today ($75-100 each). *Dolls from the collection of Marcie Tubbs. Photograph by Bob Tubbs.*

Irene Miniatures Co., Ltd.

Irene Miniatures Co., Ltd. located in Chicago, marketed a set of plastic dollhouse furniture in 1977 which used many of the early designs from the Renwal products. The "Renwal" name was obliterated from the bottoms of the furniture but the numbers from the Renwal products were still visible. "U.S.A." was also removed from the furniture as the plastic items were made in Hong Kong.

Boxes of furniture included pieces for a bedroom, living room, dining room, bathroom, kitchen, nursery, and laundry. Even extra items like Renwal's folding card table and chairs were reproduced. Because the original Renwal name has been removed, it is easy for collectors to identify these "Irene" copies. The more recent furniture was also made more cheaply than the original items. Like Renwal, the furniture is 3/4" to one foot in scale.

Pictured is a boxed set of "Irene Doll House Furniture." Many of the pieces were made with molds which had earlier been used by Renwal. The plastic furniture was 3/4" to one foot in scale. The furniture was made in Hong Kong (Boxed sets $75-100). *Furniture from the collection of Mary Lu Trowbridge. Photograph by Bob Trowbridge.*

228 Plastic Furniture and Metal Houses

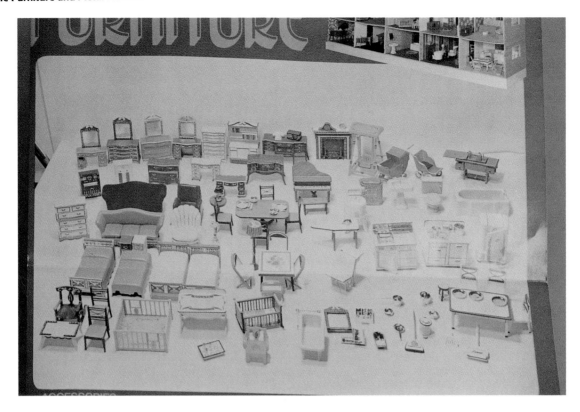

This picture from inside the box shows all the furniture made in the set of "Irene Doll House Furniture." *Furniture from the collection of Mary Lu Trowbridge. Photograph by Bob Trowbridge.*

Many of the pieces from the Irene bedroom set are similar to the earlier Renwal plastic furniture (Boxed set $75-100). *Furniture from the collection of Mary Lu Trowbridge. Photograph by Bob Trowbridge.*

Pictured is the bathroom set of Irene furniture from 1977 ($75-100). *Furniture from the collection of Mary Lu Trowbridge. Photograph by Bob Trowbridge.*

Plastic Furniture and Metal Houses 229

The boxed living room set of Irene furniture includes a fireplace and some of the pieces used as dining room furniture in the earlier Renwal era ($75-100). *Furniture from the collection of Mary Lu Trowbridge. Photograph by Bob Trowbridge.*

Boxed Irene dining room furniture featured several accessories (lamps, clock, telephone, and smoking stand) based on the original Renwal products ($75-100). *Furniture from the collection of Mary Lu Trowbridge. Photograph by Bob Trowbridge.*

Plastic Furniture and Metal Houses

Jaydon

Plastic dollhouse furniture marketed under the Jaydon name was one of the first sets of plastic dollhouse furniture to appear on the market. These 3/4" to one foot scaled pieces dated from the World War II era. Much of the furniture tended to warp but quite a number of pieces can still be found in good condition. Unlike other companies, Jaydon marketed much of the same furniture as both kitchen and dining room sets by producing the pieces in different colors. Some of the Jaydon sets also included paper rooms to be used with their furniture. See *American Dollhouses and Furniture From the 20th Century* for more information on the Jaydon furniture.

Plastic dollhouse furniture sold under the Jaydon name first appeared on the market during the mid 1940s. Some of the furniture was sold in boxes that included a room setting. These kitchen pieces were like the dining room furniture except made in different colors. Kitchen appliances were also produced (Boxed set $55-75). *Furniture and photograph from the collection of Patty Cooper.*

Jaydon living room furniture included a fireplace ($15-20), sofa $15-18), two chairs ($12-15 each), coffee table ($10-12), grand piano ($10-12), lamps ($10-12), and radio ($15-18). The plastic furniture was 3/4" to one foot in scale. The material warped easily so the furniture is hard to find in excellent condition. *Furniture and photograph from the collection of Roy Specht.*

Marx

Louis Marx and Co. was begun shortly after World War I by Louis Marx. The company remained in business for over fifty years. Although they had produced some dollhouses in their earlier years, their famous metal houses furnished with plastic furniture did not appear until 1949. Through the years, the company manufactured many different sizes and styles of these houses. The houses were furnished with a variety of plastic furniture in both the 1/2" to one foot scale and the 3/4" to one foot scale.

Besides the common plastic pieces the company also marketed a more expensive and finer set of 3/4" to one foot scale furniture in 1964. Sold under the trade name "Little Hostess," the new plastic furniture featured movable parts and fine craftsmanship. Furniture was made for a living room, dining room, kitchen, bathroom, and bedroom. Although the new furniture did not sell well in the toy line, it is now much in demand by collectors.

Metal dollhouses continued to be sold by Marx through the early 1970s so there are a variety of styles to interest collectors of today. See *American Dollhouses and Furniture From the 20th Century* for more information.

Louis Marx and Co. sold more metal dollhouses than any other firm during the years of popularity for these houses. The company also marketed many styles of vinyl "suitcase" houses. Pictured is one from 1965 which was furnished mostly with plastic Little Hostess furniture ($110-125 furnished). 15.5" high x 20" wide. *House and photograph from the collection of Roy Specht.*

Plastic Furniture and Metal Houses 231

The inside of the vinyl Marx house was furnished with Marx Little Hostess furniture and regular 3/4" to one foot Marx plastic patio pieces. *House and photograph from the collection of Roy Specht.*

Contemporary Marx metal house which is unusual because the roof lifts off to provide easy access (Furnished house $125-150). 9.5" high x 32" wide x 16" deep. *House from the collection of Nanci Moore. Photograph by Roy Specht.*

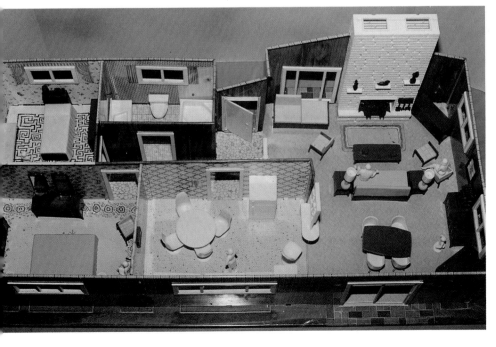

The inside of the contemporary Marx house contains five rooms furnished with modern plastic furniture. *House from the collection of Nanci Moore. Photograph by Roy Specht.*

Meritoy

Meritoy Corp. was responsible for one dollhouse now treasured by collectors. The company, based in Boston, Massachusetts, issued its only dollhouse circa 1949. The metal house featured plastic window panes as well as a chimney that went up the outside of the structure. The six-room furnished house sold for $4.98 and if purchased unfurnished, the cost was $2.98. Fifty-seven plastic pieces came with the house which included six dolls as well as the furniture. The furnished rooms were the kitchen, dining room, living room, nursery, bathroom, and bedroom. The house measured 14.5" tall x 21" wide x 10.25" deep.

Meritoy Corporation metal dollhouse dating from 1949. 14.5" high x 21" wide x 10.25" deep ($100-125). *House from the collection of Nanci Moore. Photograph by Roy Specht.*

The two-story Meritoy metal house contained six rooms and was sold either furnished or unfurnished. *House from the collection of Nanci Moore. Photograph by Roy Specht.*

Ohio Art

The Ohio Art Co., located in Bryan, Ohio, is well known for its high quality metal toys which the company manufactured over several decades. The firm, however, produced only one dollhouse. The tiny house was offered in 1949 furnished with twenty-eight pieces of plastic furniture. See *American Dollhouses and Furniture From the 20th Century* for more information.

This small metal Ohio Art Co. house contains its original plastic furniture. It is approximately 1/4" to the foot in scale (Furnished house $75-90). 5.5" high x 8.25" wide x 3" deep. *House from the collection of Nanci Moore. Photograph by Roy Specht.*

Plasco

Plasco 3/4" to one foot scaled plastic dollhouse furniture was first made by the Plastic Art Toy Corporation of America circa 1944. The company was located in East Paterson, New Jersey, and was founded by Vaughan D. Buckley. The furniture was marketed under the name "Little Homemaker" and during the late 1940s the furniture came in boxed sets with the insides of the boxes printed to represent rooms. The company made furniture to furnish a living room, dining room, kitchen, bedroom, bathroom, nursery, and garden.

The Plastic Art Toy Corporation also produced several dollhouses for their furniture. These included "Little Homemaker's Open House," a ranch house with a removable roof, and a smaller all plastic four-room house. The Plasco furniture was nicely made and many of the items included functioning drawers and doors. See *American Dollhouses and Furniture From the 20th Century* for more information on Plasco furniture and houses.

This unusual "Little Homemaker's Open House" was marketed by the Plasco Art Toy Corporation circa 1948. The house was designed with a "see through" look so that the furnishings could be viewed through the plastic as well as the open spaces ($450-500 unfurnished). 14" high x 21.5" in diameter. *House and photograph from the collection of Roy Specht.*

Four of the five rooms can be seen from the back of the open-style Plasco house. A hall makes up the sixth section of the dollhouse. *House and photograph from the collection of Roy Specht.*

Ralston Industries

Ralston Industries, located in Seattle, Washington, was responsible for marketing an unusual dollhouse in 1950. It was a circular dollhouse called "The Rallhouse" which was built on a turntable so that several children could play with it at once. There were six rooms in the house which measured 19" high by 63" around. The ceilings were 7" tall. The chimney could be used as a handle to lift the house. The Rallhouse came with 40 pieces of plastic furniture (mostly Ideal and Renwal). The house retailed for $19.95.

This Rallhouse Dollhouse was first introduced in 1950 by Ralston Industries of Seattle, Washington. It was made of fiberboard with wood details ($300 and up). 19" to top of chimney, 20" in diameter. *House from the collection of Marcie Tubbs. Photograph by Bob Tubbs.*

234 Plastic Furniture and Metal Houses

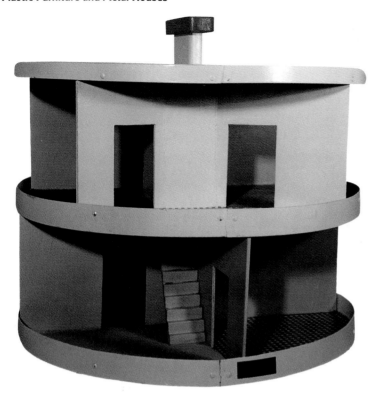

The Rallhouse was packaged with 40 pieces of furniture which appeared to be a combination of Renwal and Ideal. Some of the floors were flocked, while others were screen printed with tile. There was a stairway between floors. It was packaged in a large hexagonal "hatbox" which protected it after play. *House from the collection of Marcie Tubbs. Photograph by Bob Tubbs.*

Renwal

The 3/4" to the foot scaled Renwal plastic dollhouse furniture is probably the most well known of the early plastic dollhouse furniture. The furniture was produced by the Renwal Manufacturing Co. of Mineola, Long Island, New York. The furniture was first marketed in 1946 under the trade name "Jolly Twins." The boxes contained printed inserts that could be used as rooms for the furniture. Furniture was manufactured by Renwal to provide furnishings for a kitchen, dining room, living room, bedroom, nursery, bathroom, and laundry room.

Besides the basic rooms of furniture, Renwal also produced several items that could be purchased separately as well as different boxed sets that could add to a Renwal collection. These included a Hospital Nursery set (two sizes), Little Red School, and a veterinary set. A family of Renwal dolls as well as Renwal babies were also made to accompany the furniture. See *American Dollhouses and Furniture From the 20th Century* for more information on Renwal furniture and accessories.

Pictured is the Renwal nursery in the early trapezoidal shaped room setting. The "Jolly Twins Nursery" included a nightstand, lamp, cradle, baby buggy, bassinet, playpen, highchair, and chest of drawers. The room is 7" tall x 15" wide (at the front) x 7.5" deep (Boxed nursery $175-200). *Room and photograph from the collection of Patty Cooper.*

Twink-L-Toy

Mattel Creations, located in Los Angeles, marketed a set of Twink-L-Toy furniture in 1945 made of Plexiglas. The material was quite popular during this time period for making modern picture frames and other gift products. In an ad for the new miniatures in *Playthings* for November 1945, the company urged gift departments to carry the product so the items may not have been meant for dollhouses but instead were to be placed on shelves or table tops.

Furniture was produced for a dinette, two different living rooms, two bedrooms, and dining room. A piano and bench were sold separately.

Mattel Creations also offered a line of smaller furniture made of Plexiglas. It was approximately 3/4" to one foot in scale. Sets included a bedroom, living room, banquet room, and an extra piano. There were also display boxes made to house this furniture.

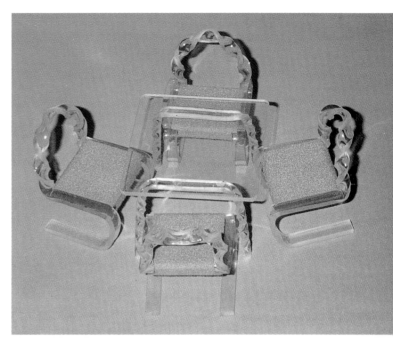

Twink-L-Toy dinette set made of Plexiglas ($75-85 boxed set). *Furniture from the collection of Becky Norris. Photograph by Don Norris.*

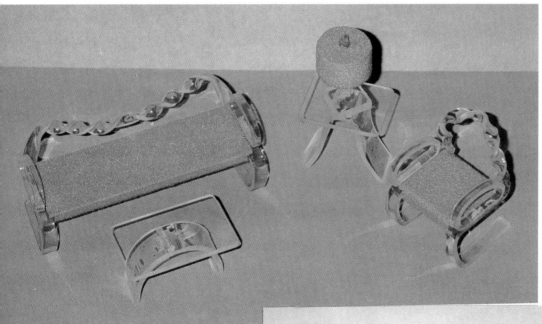

Twink-L-Toy living room also circa 1945 ($75-85 boxed set). *Furniture from the collection of Becky Norris. Photograph by Don Norris.*

Mattel Creations marketed Twink-L-Toy furniture made of Plexiglas in 1945 in a large 1" to one foot scale. Pictured is the bedroom ($75-85 for boxed set). *Furniture from the collection of Becky Norris. Photograph by Don Norris.*

236 Plastic Furniture and Metal Houses

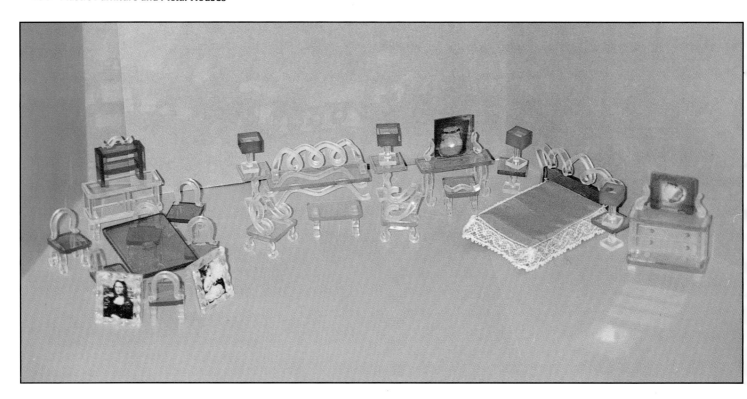

Mattel also produced a smaller scale set of Plexiglas furniture which includes pieces for a bedroom, living room, and dining room. A special box could also be purchased to display this furniture in a room setting. This furniture was marketed as "Futureland." The furniture is a small 3/4" to one foot in scale (Each room $75-85). *Furniture from the collection of Becky Norris. Photograph by Don Norris.*

Miscellaneous

In addition to all of these firms dealing with plastic furniture and metal houses, many other companies competed for the lucrative market. These included: Wolverine (Today's Kids), Multiple Products Corporation, T. Cohn, Jayline Toys, National Can Corporation, Irwin Corporation, Allied Molding Corporation, Banner Plastics Corp, Thomas Manufacturing Corp., and Best Plastics Corp., (see *American Dollhouses and Furniture From the 20th Century* for more information on these companies).

Besides these well known firms, other lesser known companies also made plastic dollhouse products. The Amloid Corp. of New Jersey produced a small all plastic house complete with furniture and Miner Industries, Inc. made several rooms furnished with plastic and wood furniture during the 1960s.

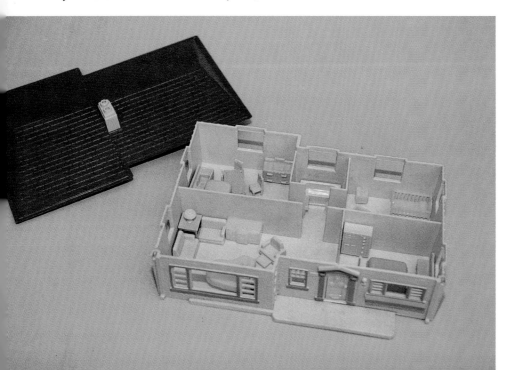

Amloid "My Dolly's House" was produced by the Amloid Corp. of New Jersey. The five room all plastic house was in the same scale as the Plasticville O-S houses and buildings of the era. It was packaged with 30 pieces of furniture and included a lift off roof. The company is still in business and is located in Saddlebrook, New Jersey ($60-75). *House from the collection of Marcie Tubbs. Photograph by Bob Tubbs.*

MISCELLANEOUS

ACCESSORIES

Although dollhouses and furniture are the mainstays for collectors, a dollhouse looks more realistic with accessories. Many companies all over the world have been involved with the production of dollhouse accessories.

German Accessories

German firms have been busy producing dollhouse accessories for more than a century. The firm of Schweizer of Diessen am Ammersee (Babette Schweizer) made metal picture frames, bird cages, music stands, mirrors, and silverware for many years. Even in later years, the Shackman Company in New York imported accessories of this type which had been made in Germany. During the later part of the 1800s, lovely wooden painted dishes were manufactured by A. Lerch in Germany. Other German companies were also producing fine china for doll houses during this same time period. Several firms exported pewter accessories including dishes, candle holders, pitchers, trays, and other serving pieces. Other necessary accessories including telephones, vacuums, clocks, radios, and typewriters were also made in Germany and exported to the United States.

German dollhouse pictures, wall sconces, and a mirror, all in the 1" to one foot scale. Circa 1900 ($20 and up each). *Accessories from the collection of Gail and Ray Carey. Photograph by Gail Carey.*

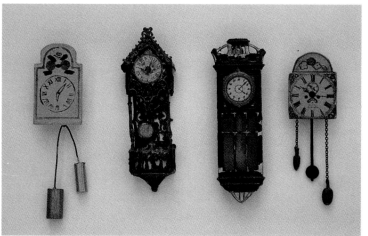

German clocks (one on right sold by Tynietoy) in the 1" to one foot scale. Circa 1900-1930s ($50 and up). *Clocks from the collection of Gail and Ray Carey. Photograph by Gail Carey.*

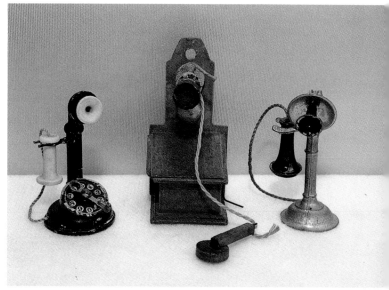

Three 1" to one foot scale German telephones in different styles ($45 and up each). *From the collection of Gail and Ray Carey. Photograph by Gail Carey.*

German bird cages (except the one on the left which is unknown). The center cage contains a wax parrot (Large $50 and up, small $20-25). *Cages from the collection of Gail and Ray Carey. Photograph by Gail Carey.*

Marked German typewriters ($40-50 each), wood "Red Line" radio ($25-35), early radio and earphones ($65-75), and horn for phonograph ($20-25). *Accessories from the collection of Gail and Ray Carey. Photograph by Gail Carey.*

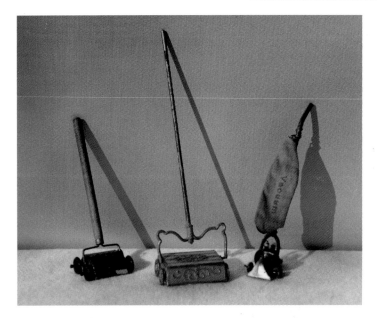

Two carpet sweepers ($40-50) and a vacuum cleaner ($60-70) all marked "Germany." *From the collection of Gail and Ray Carey. Photograph by Gail Carey.*

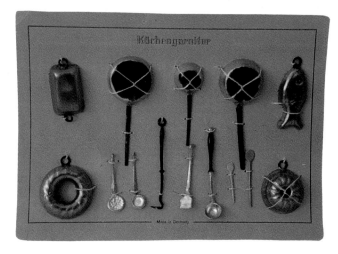

German kitchen set on original card ($35 and up). *From the collection of Gail and Ray Carey. Photograph by Gail Carey.*

German 1" to one foot scale accessories include a toast rack ($20-25), flour holder ($25-30), chafing dish ($25-30), toaster ($25-30), and brass pans ($15 a set). *Accessories from the collection of Gail and Ray Carey. Photograph by Gail Carey.*

Older German silverware and tray on original card ($25-30). *From the collection of Gail and Ray Carey. Photograph by Gail Carey.*

German 1" scale metal accessories include an electric chafing dish ($25-30), electric iron ($25-30), and a metal coffee service ($45-50). *From the collection of Gail and Ray Carey. Photograph by Gail Carey.*

German 1" scale hot plates (single burner $15-20, two burner $40-45). *From the collection of Gail and Ray Carey. Photograph by Gail Carey.*

German Christmas tree in the 1" scale ($25-30). *From the collection of Gail and Ray Carey. Photograph by Gail Carey.*

240 Miscellaneous

A collection of German metal dishes and accessories in the 1" scale ($20-50 each). These are sometimes called German pewter accessories. *From the collection of Gail and Ray Carey. Photograph by Gail Carey.*

Marked German accessories include a dinner gong ($35 and up), toast rack ($25-30), tray ($20-25), glasses and holder ($35-40), and jelly jar with spoon ($40-50). *From the collection of Gail and Ray Carey. Photograph by Gail Carey.*

Set of Bavarian china in the 1" scale ($100 and up). *Dishes and photograph from the collection of Roy Specht.*

English Accessories
See English Dollhouses and Furniture chapter

Dolly Dear

A well known early dollhouse accessory company in the United States was Dolly Dear. The business was begun in 1928 by Rossie Kirkland from Union City, Tennessee. It was first called R.T. Kirkland Co. but later became known as Dolly Dear Accessories. At first the firm was housed in Mrs. Kirkland's home but by 1943 the orders for accessories had become so large that the business was moved to its own building. This was largely due to orders from Montgomery Ward when they included the miniatures in their mail order Christmas catalog. During the war, Mrs. Kirkland's son, Carlyle, was drafted and he could no longer help in the business. Alberta Kitchell (Mrs. Kirkland's niece) took over the business end of Dolly Dear at this time. She continued to run the firm until after Mrs. Kirkland died in 1948. Alberta Kitchell Allen purchased the business from her uncle in 1950 and moved it to Rives, Tennessee.

The first illustrated catalog for Dolly Dear Accessories was issued in 1958 so it is difficult to identify the earlier items unless they are still packaged. Many of the products were carried in stock for many years with no changes in the design.

The Dolly Dear accessories were scaled 1" to one foot. Some of the items offered included: Christmas tree and stocking, vacuum cleaner, rugs, curtains, vases, bookends, mirrors, pictures, Bibles, food, clocks, telephone, typewriter, candlesticks, electric fan, toys, towel set, lamps, and a tea set.

The Montgomery Ward Christmas catalog for 1943 featured several different sets of Dolly Dear dollhouse "extras." The set of kitchen accessories sold for sixty-nine cents and included a floor mop, hand mop, wood clock, cookie jar, sugar, flour, and salt cannisters. The pantry set consisted of wood and cardboard reproductions of flour, eggs, ketchup, Maxwell House Coffee, Kelloggs' Corn Flakes, and canned food. This set also retailed for sixty-nine cents. The set of bathroom accessories included a towel, wash cloth, chenille bathmat, towel rack, shower curtain, toilet tissue, and soap. This set cost eighty-nine cents. The bedroom package consisted of a mattress, pillow, bedspread, curtain, curtain pole, and cardboard window and lamp at a cost of eighty-nine cents. The living room accessories were more expensive costing $1.35. Included were a picture, Bible, mantel decorations, waste basket, magazine rack, and rug. A 4 1/2" long rug could also be purchased separately for fifty-seven cents. Other items offered included a turkey dinner and a dollhouse luncheon set.

Although the Dolly Dear dollhouse accessories were made for over thirty years, the small items are hard to find for today's collector. The company continued in business until 1961.

The early Dolly Dear Accessories were marketed by Rossie Kirkland under the R.T. Kirkland name. Pictured are items for the dinner table (missing the cherry pie) from the 1940s ($35-40). Most of the Dolly Dear items are in the 1" to one foot scale. *From the collection of Jill H. Ramsey. Photograph by Gail Carey.*

Another set of 1940s Dolly Dear Accessories consisted of a table cover and napkins ($15-20). Also shown are two pictures made by the company ($7-10 each). *Accessories and photograph from the collection of Lois L. Freeman.*

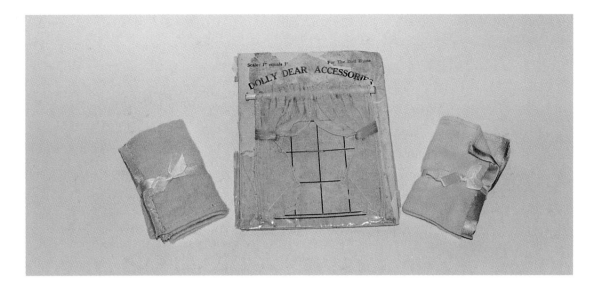

Dolly Dear provided dollhouse owners with curtains and bed clothes as well as the more usual accessories. These curtains ($15-20) date from the early period of production although the blankets ($7-10) may be from the 1950s. *Accessories and photograph from the collection of Lois L. Freeman.*

This card of Dolly Dear kitchen accessories is pictured in the 1958 catalog. The package includes flour, rolling pin, cutting board, bowl, cup, milk, cook book, spoon, and skillet ($30-35). *Items and photograph from the collection of Patty Cooper.*

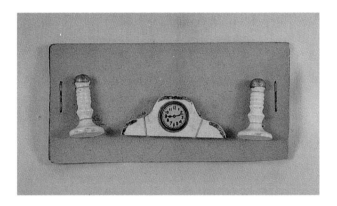

This mantel set of clock and candlesticks probably dates from the 1940s but Dolly Dear was still selling similar sets during the 1950s ($15-20). *Accessories and photograph from the collection of Patty Cooper.*

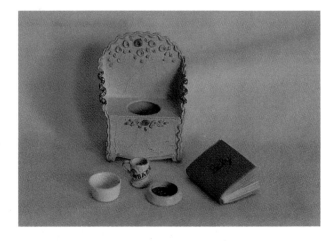

Many of the Dolly Dear accessories had a "decorated cake" look about them. Pictured is a baby potty chair ($15-18) with this design. Also shown are other baby items made by the company ($3-5 each). *Products and photograph from the collection of Patty Cooper.*

Wood lamps in the 1" to one foot scale ($8-15 depending on size). Most of these lamps are pictured in the Dolly Dear catalog issued in 1958 after the company was taken over by Alberta K. Allen. The Victorian lamp is missing its chimney. *Lamps and photograph from the collection of Lois L. Freeman.*

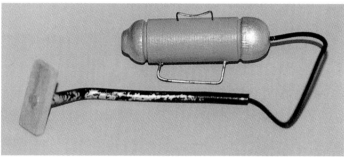

These pictures, waste baskets, and magazine racks were sold by Dolly Dear during most of their years of business ($5-10 each). The magazine rack sold for 59 cents in the 1950s. *Accessories and photograph from the collection of Patty Cooper.*

This wood vacuum sweeper was a product of Dolly Dear in the 1950s ($20-25). The company also sold typewriters, blackboards, and bookcases at this time. *From the collection of Dian Zillner. Photograph by Suzanne Silverthorn.*

Besides food on plates, Dolly Dear marketed wood canned goods as well as ketchup, eggs, and other food products ($4-6 each). *Food and photograph from the collection of Patty Cooper.*

244 Miscellaneous

This Dolly Dear clock and dog were sold in the Mark Farmer catalog in the early 1960s ($8-12 each). *Accessories and photograph from the collection of Lois L. Freeman.*

Miniature Toy Company

A company, based in Chicago, called the Miniature Toy Company competed with both Dolly Dear and Grandmother Stover's during the 1940s and 1950s. Their miniatures were also in the 1" to 1' scale. The company was run by Virginia Becic. The motto for the small firm was "The Big House of Little Things."

One of the firm's most popular products was their dollhouse sterling silver silverware. Other items of interest include a kitchen set of knives, choppers, and spatula, and another set featuring a bread board, frying pan, and egg. Appliances made by the company include a vacuum cleaner with the firm's name on the bag, an electric iron and ironing board, and a portable sewing machine.

Pictured is a packaged Dolly Dear accessory sold after the company was purchased by Alberta K. Allen ($8-10). The box includes the Allen name as well as the new Rives, Tennessee, address for the company. The box contains a chenille rug which originally sold for 20 cents in the 1950s. *From the collection of Dian Zillner. Photograph by Suzanne Silverthorn.*

A boxed MinToy De Luxe Carving Set made by the Miniature Toy Co. circa 1950s (set $25-35). *From the collection of Gail Carey. Photograph by Suzanne Silverthorn.*

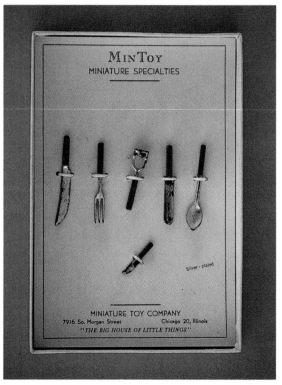

Pictured is a card of kitchen miniatures made by the Miniature Toy Company in Chicago circa 1960 (set $25-35). *From the collection of Zelma Fink. Photograph by Suzanne Silverthorn.*

Grandmother Stover's, Inc.

Perhaps one of the most famous producers of accessories was the company called Grandmother Stover's, Inc. The firm began during the early 1940s when Ohio farmer John Stover couldn't find accessories for his daughter's dollhouse and he made some himself. He met with such success that he began a business to produce miniatures to sell to other dollhouse enthusiasts. At the height of the company's sales, it made seventy-five million miniatures a year. During its many years in operation, the Stover company featured over 450 different items in its catalogs and advertisements.

The small accessories were sold as toys, dollhouse miniatures, as party favors, and to adult collectors. Several of the metal miniatures were made from the molds used by the Dowst Co. for Cracker Jack Prizes. These included a flat iron, old shoe, top hat, dinner knife, saucer, dinner plate, fish, teapot, ladder, wheelbarrow, battleship, car, and airplane.

Other very popular Stover items were the miniature copy of *The New York Times*, a deck of playing cards, Amy Vanderbilt's Etiquette book, and replicas of popular products like Drano, Kellogg's Corn Flakes, Kleenex, Ivory Soap, and Duz. Household appliances were also carried by the Stover firm to make a dollhouse complete. These included a vacuum cleaner, radio, electric iron, mixer and bowl, coffee maker, phonograph, and later a television was included in the line. In the 1970s, accessories were packaged in several different ways. A customer could purchase a small package of accessories or buy a grouping of items boxed together.

Some of the later Stover items were based on products originally made by the Miniature Toy Company of Chicago. Because that firm was no longer in business, John Stover received permission to produce several of the kitchen products.

Other items still being carried by the firm as late as 1977 included the following: metal scissors, wood rolling pin, wood milk bottle, wood bowl, wood vase, plastic baby bottle, plastic hot water bottle, plastic camera, plastic hot dog, plastic hamburger, plastic pots and pans, plastic groceries, plastic dust pan, plastic guns, plastic football, plastic music instrument, plastic lamp, plastic dice, phone book, baby book, and cook book. The ad which featured all of these items stated that the Columbus, Ohio, based company had been in business for thirty-seven years at that time. In an advertisement appearing in 1978, the company address had changed to Palm Beach, Florida.

Perhaps part of the success of Grandmother Stover's longevity was the company's ability to adjust to a changing market with the coming of television. By the late 1950s, little girls no longer seemed to be interested in dollhouse accessories so the company included items in their listing that could be used for party favors. Another factor was the economical price of the Stover items which make any dollhouse fancier of today yearn for the "good old days" when a miniature wooden rolling pin could be purchased for only ten cents.

A set of metal dishes from Grandmother Stover circa 1960s ($10-15). The company address is listed as Columbus 8, Ohio. All of the Stover accessories were in the 1" to one foot scale. *Dishes and photograph from the collection of Patty Cooper.*

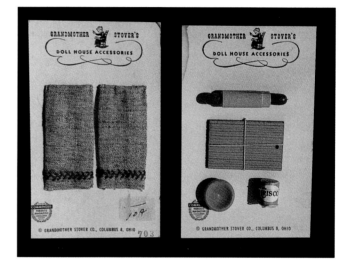

Two cards of Grandmother Stover's accessories which once sold for ten cents each. Circa 1960s ($10-15). *Accessories from the collection of Ray and Gail Carey. Photograph by Gail Carey.*

Grandmother Stover's vacuum sweeper ($10-12) and scale ($5-8). *Items and photograph from the collection of Patty Cooper.*

Miscellaneous

A large boxed set of Grandmother Stover's bathroom accessories circa late 1960s. Some of the unusual items include a toilet brush, stool lid cover, and shower curtain (Boxed set $75-80). *Set and photograph from the collection of Patty Cooper.*

The Stover accessories were sold in many different sized packages during the 1970s ($5-8 each). Included were these small cards which featured a carving set that looks much like the ones produced earlier by the Miniature Toy Company of Chicago. *Items from the collection of Dian Zillner. Photograph by Suzanne Silverthorn.*

Home decorating products were also needed by dollhouse owners during the 1970s. Grandmother Stover's accessories included pictures, clocks, mirrors, and plaques to decorate the dollhouse walls ($5-8 each). *Items from the collection of Dian Zillner. Photograph by Suzanne Silverthorn.*

Miscellaneous 247

Besides food and dishes, the Stover company produced mixers, typewriters, plants, and bathroom accessories during the 1970s ($10 each). *Accessories and photograph from the collection of Patty Cooper.*

Other popular products made by Stover were items for the pantry including canned goods and breakfast cereal ($5-8 each). *Accessories from the collection of Dian Zillner. Photograph by Suzanne Silverthorn.*

Silverware and other tableware were always good sellers for the Stover company (Boxed set $25). *Set from the collection of Dian Zillner. Photograph by Suzanne Silverthorn.*

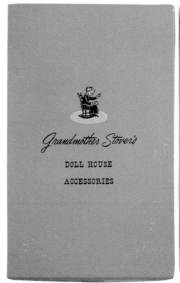

248 Miscellaneous

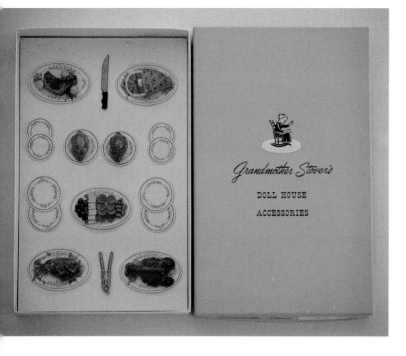

These nicely packaged boxed sets of Stover products are also circa 1970s. This one adds dishes and food to a table setting (Boxed set $25). *Box from the collection of Dian Zillner. Photograph by Suzanne Silverthorn.*

This kitchen set included pots and pans as well as kitchen utensils for the dollhouses of the 1970s (Set $25). *Accessories from the collection of Dian Zillner. Photograph by Suzanne Silverthorn.*

A Stover metal tea set from the 1970s included candlesticks and a cookie plate as well as the more ordinary pieces (Set $25). *Accessories from the collection of Dian Zillner. Photograph by Suzanne Silverthorn.*

Grandmother Stover's kept up with the times by offering a television as an accessory during the company's later years ($10-15). *From the collection of Gail and Ray Carey. Photograph by Gail Carey.*

A more varied set of Stover accessories included a telephone, phone book, cards, checkers, a record player and records, a door knocker, mailbox, and porch lights (Set $25). *From the collection of Dian Zillner. Photograph by Suzanne Silverthorn.*

Miscellaneous Accessories

As people continue to hunt for the elusive older accessories, their scarcity sometimes leads dollhouse enthusiasts to improvise. Some collectors try the "do it yourself" method while others convert practical products into dollhouse accessories. Items that are included in this category are pencil sharpeners and salt and pepper shakers that are shaped like clocks, phonographs, globes, television sets, and lamps. Some party favors, gum ball machine prizes, and charms also can be used as accessories. Even jewelry can be modified into a light fixture or the charms from a bracelet can add groceries to a kitchen or a toy to a child's room.

Dollhouse rugs can also be made from practical products. Small doilies or pot holders can be pressed into service as dollhouse floor coverings. Collectors can also make rugs from bits of paisley scarves or pieces of felt. One of the most popular sources for dollhouse rugs are the premiums given many years ago with the purchase of tobacco products. The rugs came in several sizes from 3" x 5" to 8.5" x 12." These rugs give the collector a size for almost any dollhouse floor.

With the expansion of the sale of new dollhouse kits during the 1980s and the 1990s, dollhouse accessories have once again become big business. Even today's collectors of older dollhouses and furniture often have to include some of the new accessories in their collections to make houses complete. With the older accessories so hard to find, new pictures, mirrors, pots, pans, and dishes can be added to the old dollhouses while the collector continues the hunt for older, more appropriate accessories to compliment a dollhouse collection.

See Tynietoy and Tootsietoy for accessories carried by those companies.

Solid brass fireplace set in 3/4" scale made by the Lee Manufacturing Co. in Hartford, Connecticut ($20-25 set). *From the collection of Gail and Ray Carey. Photograph by Gail Carey.*

Garden tools on original card marked "Made in U.S.A/B.S.&Co" ($30-40). Also pictured are potted plants by an unknown maker ($15-20 each). *From the collection of Gail and Ray Carey. Photograph by Gail Carey.*

Desk set made by an unknown company, possibly Dolly Dear. It is in the 1" to one foot scale and is circa 1950 (Set $40 and up). *Set and photograph from the collection of Lois L. Freeman.*

The tin trunk on the left was made by Marx ($30-40). The one on the right is unmarked ($20-25). *From the collection of Gail and Ray Carey. Photograph by Gail Carey.*

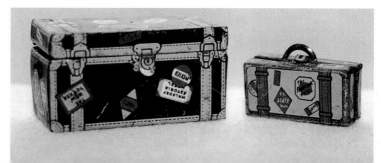

250 Miscellaneous

Practical products like salt and pepper shakers (lamp), pencil sharpeners (clock), gum ball charms, and jewelry can be used as dollhouse accessories by collectors. *From the collection of Dian Zillner. Photograph by Suzanne Silverthorn.*

Tobacco rugs are very popular with today's dollhouse collectors ($5-15 each). They were made in several different sizes from 3" x 5" to 8.5" x 12". The rugs were made as premiums and were sold with the tobacco products. *Rugs and photograph from the collection of Patty Cooper.*

Many different patterns were made in the tobacco rugs. These feature Native American designs that fit nicely in the lithographed cabins produced by Bliss and other firms in the early part of the century. *Rugs and photograph from the collection of Patty Cooper.*

DOLLHOUSE DOLLS

Bisque Dolls

Families of dollhouse dolls have been commercially made since the late 1870s. Before this time most of the dolls used in dollhouses were the early wood Peggity dolls or dolls with china heads. The early dollhouse dolls were made with bisque heads, as well as bisque lower arms and legs, and cloth bodies. Usually the hair was molded and the eyes painted but some models included glass eyes and/or wigs. These dolls were made in many images including mother, father, sister, brother, grandmother, grandfather, cook, maid, butler, nurse, and chauffeur. Some of the men dolls wore mustaches or beards, and/or molded hats. In 1913 the 5" to 7" dolls could be purchased for 50 cents to $1.50 each. Today, the most expensive of these dolls are the men with molded hats, the women with glass eyes, and the dolls made to represent black servants. One of the nicest of the women dolls is the model produced by the German firm of Simon and Halbig complete with glass eyes and a wig. Collectors call this doll a "Little Women" type. It was made circa 1900 and ranges in size from 7" to 9" tall with the mold mark #1160.

Although these early bisque dollhouse dolls were made by companies in both France and Germany, the ones from Germany are most often found. Manufacturers making these dollhouse dolls include: Heinrich Schmuckler, Schindhelm and Knauer, Friedmann & Ohnstein, and Welsch and Co. Similar dolls were still being produced during the 1920s. The German dolls from this period can be recognized by their later style molded shoes and the short bobbed hairstyles of the women. If the dolls are in their original clothing, the women's dresses are shorter and more modern than their earlier counterparts.

In addition to the special dollhouse dolls, manufacturers produced many other styles of all bisque dolls that can represent children and babies in an older dollhouse. The best of these images for dollhouse use are the small dolls under 5" tall that have jointed arms and legs. Some of the dolls featured wigs and glass eyes while most were produced with molded hair and painted eyes. The tiny dolls of this type, ranging from 2" to 3" tall, are small enough that they can be used with the 1/2" to one foot scale of dollhouse furniture. Some of these dolls were made with molded hats indicating either a male or a female doll. Most of these dolls were also produced in Germany.

Dollhouse dolls with bisque heads, lower arms and legs, and cloth bodies. The man has a mustache and both dolls have molded hair. The lady is 6" tall ($140-160) while the man is 6.25" in height ($195-210). Circa 1900-1910, made in Germany. *From the collection of Gail and Ray Carey. Photograph by Gail Carey.*

Beautiful German dollhouse doll with bisque head, wig, full bisque arms, and painted features ($250-275). She is 7" tall and is wearing what appears to be her original clothing circa 1900-1910. *From the collection of Gail and Ray Carey. Photograph by Gail Carey.*

A group of bisque dollhouse dolls ranging in size from 4.75" to 5" tall. Most are wearing their original clothing. All have bisque heads, lower arms and legs, and painted features. The cook has a mustache ($195-210). Circa 1900-1910 except for the two men who may date from 1920 (Other dolls $165-185). All made in Germany. *From the collection of Gail and Ray Carey. Photograph by Gail Carey.*

252 Miscellaneous

This group of bisque dollhouse dolls features a maid with a molded cap and a man with a bald head ($200 and up each). Most are wearing their original clothing. The dolls range in size from 4" to 5.75". Although the man appears older, the other dolls are circa 1920 and all were probably made in Germany ($135-150 each). *From the collection of Gail and Ray Carey. Photograph by Gail Carey.*

These dolls suitable for dollhouses include a man in uniform ($150 and up), a bride and groom and a girl all measuring 4" to 4.25" in height ($125 and up each). The dolls are circa 1910-1920 and were probably all made in Germany. *From the collection of Gail and Ray Carey. Photograph by Gail Carey.*

This grouping of dolls includes one dollhouse lady 5" tall ($175-200), a small all bisque doll with jointed arms measuring 2.75" ($50-60), and an unusual all bisque doll with glass eyes, mohair wig, and a swivel neck ($125-150). She is 4.25" in height. All are German dolls circa 1910-1925. *From the collection of Gail and Ray Carey. Photograph by Gail Carey.*

Bisque shoulder head dollhouse dolls all dating from 1890 to 1915. They range in height from 5" to 6". All have cloth bodies and lower bisque arms and legs. The largest doll in the center was made by Simon & Halbig and is marked "S&H" on her shoulders ($300-350). This doll has glass eyes, and a mohair wig and is called a "Little Women" type by collectors. The doll on the left represents a motorist with his molded hat and goggles ($250-300). Dolls probably all from Germany. *Dolls and photograph from the collection of Linda Hanlon.*

Bisque shoulder head dolls with cloth bodies and bisque arms and legs. The dolls range in size from 5" to 7" tall. The unusual black butler wears his original clothing ($400 and up). The dolls are probably all from Germany (Other dolls $150-200). *Dolls and photograph from the collection of Linda Hanlon.*

This rare 7" tall soldier with molded hat and uniform was also made in Germany. He has a bisque shoulder head, painted features (including a mustache), lower bisque arms and legs and a cloth body. Circa 1900 (Not enough examples to determine a price). *Doll and photograph from the collection of Anne B. Timpson.*

These bisque dollhouse dolls are dressed for a game of tennis. They are 6.5" tall. The lady has glass eyes and a wig and is in the style of the Simon & Halbig dolls. The man's features are painted and he has a mustache. Both dolls have cloth bodies with bisque arms and legs and were made in Germany (Man $250 and up, woman $350 and up). *Dolls and photograph from the collection of Anne B. Timpson.*

This lovely bisque doll is 6.5" tall and is also in the style of the Simon & Halbig dolls with glass eyes and her original wig. She has a bisque shoulder head, full bisque arms and bisque legs ($350 and up). *Doll and photograph from the collection of Anne B. Timpson.*

By the 1920s, the dollhouse dolls became more modern looking. The women had bobbed hair, the men no longer wore beards or mustaches, and the dresses on the women were much shorter. Pictured are two of these types of dolls. They still were made with bisque shoulder heads, cloth bodies, and bisque lower arms and legs. This pair measures 5.25" tall and are also of German origin (Man $135-150, woman $100-125). *Dolls from the collection of Gail and Ray Carey. Photograph by Gail Carey.*

256 Miscellaneous

This group of dollhouse dolls in a smaller size also dates from the 1920s. These dolls are approximately 2.25" tall, wear their original costumes, and are all made of bisque. The maid's cap is molded on her head (Maid $125 and up, other dolls $50-75). The dolls are all from Germany. *Dolls and photograph from the collection of Patty Cooper.*

Four more German dolls from the 1920s are pictured. The dolls range in size from 3" to 5" tall. Three of the dolls are wearing their original clothing ($95-110 each), one has glass eyes ($135-175), and two feature wigs ($125 and up). *Dolls from the collection of Dian Zillner. Photograph by Suzanne Silverthorn.*

Another example of the all bisque dolls of the 1920s. This family of dolls ranges in size from 2 1/2" to 4" tall. They wear molded hats ($75 and up each) and all the examples are from Germany. *Dolls and photograph from the collection of Patty Cooper.*

A group of small all bisque dolls that can be used as children in dollhouses. The baby is 1.5" tall ($40-60) and the others range in size from 2.75" to 3.75" in height ($75-150 each). Glass eyes and wigs are featured on several of these German dolls circa 1900-1920. *Dolls from the collection of Gail and Ray Carey. Photograph by Gail Carey.*

German bisque babies also add an interesting element to a dollhouse when a buggy or crib is present. Pictured are two that measure 2.5" tall on the left ($55 and up each) and 1.75" tall on the right ($40 and up). Circa 1910-1925. *Dolls from the collection of Gail and Ray Carey. Photograph by Gail Carey.*

China Dolls

Although the small dolls made of china are not as popular for dollhouse dolls as the more attractive bisque, these dolls have made many dollhouses "home." The most prolific of these dolls are those with the "common" hair style. This model was manufactured in Germany from 1900 to the beginning of World War II. The dolls can be dressed as either men or women which adds to their appeal. The heads, lower arms, and lower legs on these dolls were made of china while the bodies were cloth. "Frozen Charlotte" china dolls have also been used as dollhouse dolls, over the years, but their inability to bend gives them a disadvantage not shared by the china dolls with the cloth bodies.

German china dolls were used for dollhouse inhabitants before the German bisque dolls were produced. China head dolls can be found in sizes as small as 3" tall. The smallest "Frozen Charlotte" dolls (unjointed china) were only 1" tall. Pictured are two dolls with china shoulder heads, lower china arms and legs, and cloth bodies. These common hair chinas could be dressed as either men or women. Also shown is a "Frozen Charlotte" doll. The dolls (circa 1910-1930) range in size from 5" to 7" ($45-75 each). *Dolls from the collection of Dian Zillner. Photograph by Suzanne Silverthorn.*

Painted Bisque Dolls

As the interest in the older bisque dollhouse dolls lagged in the late 1920s and early 1930s, the earlier dollhouse dolls were no longer made. Both Germany and Japan began producing small inexpensive dolls to fill the void. The new dolls were made of a cheaper painted bisque material and ranged in size from 2 1/2" to 7" tall. Although these dolls are the right size for dollhouses and date from the period of Schoenhut and Strombecker furniture, they have a disadvantage because the dolls available are mainly models of children so parents are at a premium. The dolls, however, offer an inexpensive alternative for a supply of dollhouse dolls.

Small inexpensive dolls produced in Germany circa late 1920s and early 1930s. They were made of a painted bisque material. The jointed dolls range in size from 3" to 6" and have painted features and molded hair ($25-40 each). *Dolls and photograph from the collection of Patty Cooper.*

The same type of inexpensive painted bisque dolls were also made in Japan. Pictured are dolls ranging in size from 4" to 5.5". The jointed dolls have painted features, molded hair, and date from the 1930s ($20-30 each). *Dolls from the collection of Dian Zillner. Photograph by Suzanne Silverthorn.*

Grecon Dolls

Although English dollhouses have been popular for many years, most of the early dollhouse residents were imported from Germany. The famous Grecon dolls which are thought of as English products also had their beginnings in Germany. The dolls originated when Margarete Cohn started making dolls in Berlin in 1917. In 1920 she began using the trade name "Grecon" and marketed the dolls in Germany. After Cohn moved to London in 1936 she located her doll operation in that city. Margarete Cohn continued in the doll business until the 1980s after moving to Haywards Heath, Sussex, in 1959.

The Grecon Dolls were of the armature type with bodies constructed of wire frames padded with yarn. The heads featured embroidered hair and painted features. The dolls' shoes were made of metal so they were able to stand. The Grecon dolls were produced in both 3/4" and 1" to the foot scale. A variety of costumes were designed for the dolls and at one point in the firm's earlier years, fifty different dolls were offered. The Grecon dolls were sold by the English Barton firm for many years and the dolls are easily confused with the English Dol-Toi dolls which are similar. See information on these dolls in the Dol-Toi section. Other dolls which resemble the Grecon dolls were produced in Scotland and bear the trade name Tomac.

A grouping of Grecon dolls made in England. The bodies are of the armature type, with embroidered hair and painted features on the heads. The shoes are of metal. The adults are approximately 3.5" tall and are wearing their original clothing. They date from the 1950s-1970s ($30-35 each). *Dolls and photograph from the collection of Patty Cooper.*

English Grecon dolls were also produced in the 1" to one foot scale. Pictured are several from this line in their original clothing. The dolls on the far left and the far right are marked with the "Grecon" name ($40-45 each). *Dolls and photograph from the collection of Lois L. Freeman.*

260 Miscellaneous

Caco Dolls

The concept of using wrapped limbs and bodies over wire armatures was used by many companies and is still being done today. Most of the later dolls have been imported from Japan or Hong Kong and are made with plastic or vinyl heads and limbs. The nicer earlier dolls feature composition heads with painted features. Dolls of this type were produced in Germany beginning in the 1930s under the name "Caco." The early dolls had metal hands and feet and they came in many different costumes. Some of the dolls had wigs but most featured molded hair. The dolls were produced in both the 3/4" and 1" to one foot scale. Later editions of Caco dolls were made with plastic heads and hands along with the metal feet. Current dolls have plastic feet as well. The boxes for these dolls are marked "Made in Western Germany/ Gebrauchsmusterschutz." These later dolls have molded hair.

Dolls similar to Grecon were made by Tomac Toys in Scotland. These dolls are also in the 3/4" to one foot scale and are all original ($20-25 each). *Dolls from the collection of Ray and Gale Carey. Photograph by Gale Carey.*

Caco dolls (1" to one foot in scale) made in Germany circa 1930s. The dolls feature metal feet and hands and heads made of a composition type material. The bodies have wrapped limbs over wire armatures ($35-45 each in original clothing). *Dolls and photograph from the collection of Lois L. Freeman.*

Another grouping of German 1" scaled Caco dolls from the 1930s ($35-45 each in original clothing as pictured). One of these dolls wears a wig. *Dolls and photograph from the collection of Patty Cooper.*

Miscellaneous 261

Caco dollhouse family in the 1" scale dating from 1968. The dolls were then sold for $1.50 to $3.00 each and had plastic heads and hands and metal feet ($25-30 each in original clothing). *Dolls and photograph from the collection of Lois L. Freeman.*

Current Caco dollhouse dolls have plastic heads, hands, and feet, and are made with molded hair and painted features. The pictured dolls are in the 1" scale ($25-35 each). *Dolls from the collection of Dian Zillner. Photograph by Suzanne Silverthorn.*

Caco dolls were also made in the 3/4" to 1" scale. Shackman sold this boxed set in the early 1960s. The father is 3.5" tall and all the dolls have plastic heads and metal hands and feet (Boxed set $100 and up). *Dolls and photograph from the collection of Patty Cooper.*

Flagg Dolls

The 1940s brought several new concepts in the making of dollhouse dolls to accompany the new plastic dollhouse furniture. With the end of World War II, more and more companies were making use of plastics to manufacture new products.

The Flagg Doll Company of Jamaica Plain, Massachusetts, used plastic when they produced the Flagg Flexible Play Dolls. The firm began the production of these dolls around 1948. Although they used the old wire armature method of doll construction, a new innovation of adding a covering of plastic to the frame made the dolls more life-like. The dolls could still bend easily. The Flagg dolls had painted features and "real" clothes along with molded hair.

The dolls were designed by Sheila Markham Flagg in several different sizes. They were varied enough that they could be used with either the 3/4" to one foot scale of furniture or the 1" to one foot scale. The dolls in the larger scale varied in size from 6" for the father doll to 4.5" for the child dolls. In the 3/4" scale, the father was 4.5" tall and the child dolls measured 3.5" tall. Of course, the baby dolls were even smaller.

Many other models of play dolls were made besides the family dolls for dollhouses. Some of these dolls were as tall as 7". By the 1950s the firm included over forty different character dolls in its line.

In 1973 the company was sold to Ralph D. Eames and he continued the business on a lesser scale until 1985 when it was closed. The last products to be made by the Flagg company were the dollhouse family dolls.

Plastic dolls made by the Flagg Doll Company of Jamaica Plain, Massachusetts. The dolls had a covering of plastic added to the wire armature to make them more life-like. This set of boxed family dolls in the 3/4" to one foot scale dates from the late 1950s or early 1960s (Boxed set $100-125). These dolls were also made in larger sizes. *Dolls from the collection of Ray and Gail Carey. Photograph by Gail Carey.*

Twinky Dolls

As the interest in plastic products grew, more dollhouse dolls were made of this material. The Renwal, Plasco, and Ideal companies all manufactured plastic dolls to accompany their dollhouse furniture (see *American Dollhouses and Furniture From the 20th Century*).

Another very interesting family of plastic dollhouse dolls was marketed by Ethel R. Strong beginning in 1946. The dolls were produced for Mrs. Strong by Mr. Bauer from Lemonister, Maine. These "Twinky" dolls were in the 1" to one foot scale and included a man 6" tall, a woman 5.5" tall, a boy and a girl each measuring 4" tall, and a baby only 2" tall. The dolls had painted features and molded hair and had removable clothing. The most unusual aspect of the dolls was their construction. Not only were the dolls jointed at the hip and shoulder but the adults also included joints at the knees. The dolls were strung with elastic.

For a short time in 1950 the dolls were carried by the Grandmother Stover company. The dolls sold for from $1.00 (baby) to $2.25 (adults) in the Stover advertisements.

During the years the dolls were produced, Mrs. Strong marketed about seventy-five different types of costumed dolls including those representing parents, children, maids, nurses, grandparents, clowns, and historical and religious figures.

Twinky dolls marketed by Ethel R. Strong beginning in 1946. The plastic jointed dolls are in the 1" to one foot scale and the adult dolls also feature jointed knees. In 1950 the Grandmother Stover company was selling the dolls at prices from $1.00 to $2.25 each ($50 to $60 each). *Dolls and photograph from the collection of Lois L. Freeman.*

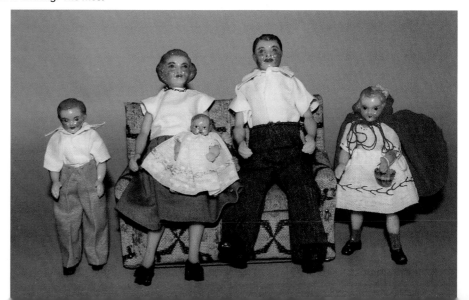

Erna Meyer Dolls

Erna Meyer, another very famous maker of dollhouse dolls, began producing dolls at about the same time as Ethel Strong. These dolls, however, were made of cloth instead of plastic. The German Erna Meyer dolls were first made in October 1945. The new dolls used the armature construction and were all handmade. The dolls were first exhibited in Nuremberg in 1950. They were meant to be 1" to one foot in scale but some of the dolls vary from that size. The dolls are still being made today and the doll packages of the more recent dolls are labeled with the "Ermey" trade name. The packages also state that the dolls are flexible and were handmade in Western Germany. The features on the dolls' faces are painted and each is fitted with a small wig. Because the dolls have been produced for decades in so many different models, they offer collectors a varied selection.

Erna Meyer cloth dolls were made in Germany beginning in 1945. These dolls are circa 1970s and are wearing their original clothing. Dolls with cardboard on their shoes are older. These dolls are in the 1" to one foot scale ($25-35 each). *Dolls and photograph from the collection of Patty Cooper.*

Recent Erna Meyer dolls have plastic feet and a more modern look ($40-45 each). *Dolls and photograph from the collection of Patty Cooper.*

264 Miscellaneous

Pictured are two early Erna Meyer dolls with more sculptured faces probably circa late 1950s. These dolls are also larger than most of the later dolls measuring 6.75" tall ($30-40 each). *Dolls and photograph from the collection of Lois L. Freeman.*

Miscellaneous Dollhouse Dolls

Besides the well known companies, many lesser known firms also produced dollhouse dolls. The plastic "Carol Lee Doll House Dolls" were made in Denver, Colorado, in a small 1/2" to one foot scale. Perhaps the dolls were to be used in the metal Marx dollhouses made in the same scale. More recently, other dolls have been produced in Hong Kong in a 3/4" to one foot scale. These dolls called "Bend-a-Family" were also made of plastic. Shackman and Mark Farmer catalogs from the 1960s and 1970s always featured dollhouse dolls for sale. Some of these dolls were from the popular firms but many were marketed by unknown companies. Some of these dolls came in kits and the dolls were to be assembled by the consumer. These are still sold today to offer collectors the alternative of making their own "antique" dollhouse dolls at reasonable prices.

Boxed set of plastic dollhouse family dolls. They are jointed at the hips so they can sit. The scale is approximately 1/2" to one foot. The box is marked "Carol Lee Doll House Dolls/Denver, Colo." Date unknown.(Boxed set $50 and up). *Dolls and photograph from the collection of Betty Nichols.*

Boxed "Bend-a Family" dolls made in Hong Kong. The family of dolls is made of flexible plastic so they can bend easily. The dolls are 3/4" to one foot in scale and probably date from the 1980s (Boxed set $25-30). *Dolls from the collection of Dian Zillner. Photograph by Suzanne Silverthorn.*

SOURCES

DEALERS

Linda Boltrek
P.O. Box 15
Captiva Islands, Florida
33954

Lisa Boutilier
91 B Brightwood Ave.
Worcester, MA 01604
508-791-4563
SASE for information on specific area of interest.

Dolls House Antiques
290 Summerhaven Drive North
East Syracuse, NY 13057
Mail order and shows.

Zelma Fink
1115 De Vere Drive
Silver Spring, MD 20903
301-434-6823
Antique and Collectible Miniatures.
SASE with your detailed inquiries.

Marilyn's Miniatures of Marshallville
Marilyn Pittman
P.O. Box 246
Marshallville, OH 44645
Dollhouse and dollhouse furniture mail order.

Judith A. Mosholder
186 Pine Springs Camp Road
Boswell, PA 15531
Send SASE for list of plastic dollhouse furniture.

Bobbie Segal
P.O. Box 39
Julian, PA 16844
814-355-2542
Send "want" list.

Linda and Carl Thomas
345 Mountainview Drive
Bluefield, WV 24701

Paige Thornton
P.O. Box 670568
Marietta, GA 30066
For "List of Lists" of collectible and antique
miniatures send SASE.

Anne B. Timpson
201-228-3230
Specializing in antique dollhouses and dollhouse furniture, accessories,
and dolls from 1820s to 1930s; also stores, stables, and theaters.

Toys in the Attic
Gaston, Joan and Renee Majeune
167 Phelps Avenue
Englewood, N.J. 07631
201-568-6745
Dollhouses, furniture, shops, iron, tin and paper lithograph toys. Shows
and photos on request.

PUBLICATIONS

Antique Doll World
I.C. Holdings Inc., 225 Main Street, Suite 300
Northport, NY 11768-1737

Dollhouse and Miniature Collector's Quarterly
Editor: Sharon Unger
P.O. Box 16
Bellaire, MI 49615

Dolls House and Miniature Scene
EMF Publishing, 7 Ferringham Lane
Ferring, West Sussex, BN12 5ND, England

Dolls House World
Ashdown Publishing Limited, Avalon Court, Star Road,
Partridge Green, West Sussex, RH13 8RY, England

International Dolls House News
Nexus Special Interests, Tower House, Sovereign Park,
Lathkill Street, Market Harborough, Leicester LE16 9EF,
England

Miniature Collector
30595 Eight Mile Rd.
Livonia, MI 48152

Nutshell News
21027 Crossroads Circle
Waukesha, WI 53187

MUSEUMS

Angels Attic
516 Colorado Avenue
Santa Monica, CA 90401
310-394-8331

Margaret Woodbury Strong Museum
One Manhattan Square
Rochester, New York 14607
716-263-2700

Toy and Miniature Museum of Kansas City
5235 Oak St.
Kansas City, MO 64112
816-333-2055

Washington Dolls' House and Toy Museum
5236 44th St. NW
Washington D.C. 20015
202-363-6400

Leslie and Joanne Payne from Dolls House Antiques sell miniatures through mail order and at shows.

Zelma Fink selling miniatures at the 1996 convention of the National Association of Miniature Enthusiasts.

Joan and Gaston Majeune from Toys in the Attic also sell dollhouses through mail order and shows in many different states.

BIBLIOGRAPHY

Ackerman, Evelyn. *Dolls in Miniature*. Annapolis, MD: Gold Horse Publishing, 1993.

_____. *The Genius of Moritz Gottschalk*. Annapolis MD: Gold Horse Publishing, 1994.

_____. *Victorian Architectural Splendor in a Nineteenth Century Toy Catalogue*. Culver City, CA.: ERA Industries, 1980.

Adams, Margaret, ed. *Collectible Dolls and Accessories of the Twenties and Thirties from Sears, Roebuck and Co*. New York: Dover Publications, 1986.

Baker, Linda. *Modern Toys: American Toys 1930-1980*. Paducah, Kentucky: Collector Books, 1985.

Bliss, R. *Fall Catalogue, 1895. R. Bliss Manufacturing Co*. Pawtucket, R.I.: R. Bliss, 1895. Photocopy courtesy of the Margaret Woodbury Strong Museum.

Bliss, R. Manufacturing Company. *1896 Fall Catalogue of the R. Bliss Manufacturing Co*. Pawtucket, R.I.: R. Bliss, 1896 reprint ed., Washington, D.C.: Antique Toy Collectors of America, Inc., 1986.

Bliss, R. *The R. Bliss Mfg. Co.'s Catalogue of Lithographed Toys, Wood Novelties, and Tool Chests, 1901*. Pawtucket, R.I.: 1901. Photocopy courtesy of the Margaret Woodbury Strong Museum.

Block House, Inc. Catalogs 1940, 1950, 1977. New York: Block House, Inc.

Boyce, Charles. *Dictionary of Furniture*. New York: Henry Holt and Company, 1985.

Chestnut Hill Studio. Catalog. Taylors, S. C., Chestnut Hill, 1951.

Coleman, Dorothy, Elizabeth, and Evelyn. *The Collector's Encyclopedia Of Dolls*. New York: Crown Publishers, Inc., 1968.

The Doll & Toy Collector. Swansea, Wales: International Collectors Publications, July/August 1983 and September/October 1983 issues.

Dollhouse and Miniature Collectors Quarterly. Advertisements reprinted in various issues 1990-1996. Bellaire, MI.

Dolly Dear Accessories. Rives, Tennessee: Dolly Dear, 1958.

Dol-Toi Products. Catalog 1964-5. Stamford, England: Dol-Toi, 1964.

Earnshaw, Nora. *Collecting Dolls Houses & Miniatures*. London: New Cavendish Books, 1993.

Eaton, Faith. *The Ultimate Dolls' House Book*. London: Dorling Kindersley, 1994.

Foulke, Jan. *12th Blue Book Dolls and Values*. Cumberland, Maryland: Hobby House Press, Inc., 1995.

Grandmother Stover's Doll House Accessories. Columbus, OH: Grandmother Stover, 1977.

Greene, Vivien. *The Vivien Greene Dolls' House Collection*. London: Cassell, 1995.

Hall's Lifetime Toys. Catalogs from the 1970s. Chattanooga, TN.

International Dolls House News. Various issues. Leicester, England: Lexus Special Interests.

Jackson, Valerie. *A Collector's Guide to Doll's Houses*. Philadelphia, PA.: Running Press, 1992.

_____. *Dollhouses The Collectors Guide*. Edison, New Jersey: Book Sales, Inc., 1994.

Jacobs, Flora Gill. *A History of Dolls' Houses*. New York: Charles Scribner's Sons, 1965.

_____. *Dolls' Houses in America*. New York: Charles Scribner's Sons, 1974.

Jendrick, Barbara Whitton. *Paper Dollhouses and Paper Dollhouse Furniture*. n.p.: By the Author, 1975.

Johnson, Audrey. *Furnishing Dolls' Houses*. Newton Centre, MA.: Charles T. Branford Co., 1972.

King, Constance Eileen. *The Collector's History of Dolls' Houses*. New York: St. Martin's Press, 1983.

Livingston, Barbara. "The Furniture of Gottschalk." *Antique Doll World*, September/October 1995, pp. 39-42.

MacLaren, Catherine B. *This Side of Yesterday in Miniature*. LaJolla, CA.: Nutshell News, 1975.

McAlester, Virginia and Lee. *A Field Guide to American Houses*. New York: Alfred A. Knopf, 1986.

McClintock, Marshall and Inez McClintock. *Toys in America*. Washington, D.C.: Public Affairs Press, 1961.

Mark Farmer Co. Inc. Catalog 1968. El Cerrito, CA: Mark Farmer Co., Inc.

Mitchell, Donald and Helene. *Dollhouses Past and Present*. Paducah, KY.: Collector Books, 1980.

Montgomery Ward. Catalogs. Various issues from 1923-1980. Chicago: Montgomery Ward.

O'Brien, Marian Maeve. *The Collector's Guide to Dollhouses and Dollhouse Miniatures*. New York: Hawthorn Books, 1974.

Osborne, Marion. *Bartons "Model Homes."* Nottingham, England: By the Author, 29 Attenborough Lane, Chilwell NG9 5JP, 1988.

_____. *Dollhouses A-Z*. Nottingham, England: By the Author, 29 Attenborough Lane, Chilwell NG9 5JP, n.d.

_____. Continuing series on Tri-ang, Amersham, Tudor Toys, and others. *Dolls House and Miniature Scene*. West Sussex, England: EMF Publications.

_____. *Lines and Tri-ang Dollhouses and Furniture 1900-1971*. Nottingham, England: By the Author, 29 Attenborough Lane, Chilwell NG9 5JP, 1986.

Pinsky, Maxine A. *Greenberg's Guide To Marx Toys*. Sykesville, MD: Greenberg Publishing Co., Inc., 1988.

Rosner, Bernard, and Jay Beckerman. *Inside the World of Miniatures & Dollhouses*. New York: Bonanza Books, 1976.

Ruddock, Pam. "Collectors' Corner: Eagle Eyed Daughter Helped to Unravel Tin Mystery." *Dolls House World*.

Schmuhl, Marian. "Pliant Playthings of the Past." *Dolls: The Collectors Magazine*. December 1993, pp. 50-56.

Schroeder, Joseph, ed. *The Wonderful World of Toys, Games and Dolls 1860-1930*. Northfield, Illinois: Digest Books, Inc., 1971.

Schwartz, Marvin. *F.A.O. Schwarz Toys Through the Years*. Garden City, New York: Doubleday and Co., Inc., 1971.

Sears, Roebuck and Company. Catalogs. Various issues from 1900-1982. Chicago: Sears, Roebuck and Company.

Smith, Patricia. *Doll Values Twelfth Edition*. Paducah, KY.: Collectors Books, 1996.

Snyder, Dee. "The Collectables." *Nutshell News*, "Dolly Dear Accessories," July-August 1979; "Toncoss Miniatures," May 1981; "Lynnfield/Block House Furniture," January 1987; "Following Lynnfield-Block House," February 1987; "Sonia Messer Imports," March 1987; "Min Toy," May 1988; "Mary Frances Line," January 1990; "Exclusive Offering," May 1990; "Colorful Canadians," November, 1990.

Sonia Messer Imports. Various catalogs circa 1980s. Los Angeles, CA.: Sonia Messer Imports.

Stevenson, Katherine Cole and H. Ward Jandi. *Houses by Mail: A Guide to Houses from Sears, Roebuck and Company*. Washington, D.C.: The Preservation Press, 1986.

Stille, Eva. *Doll Kitchens 1800-1980*. West Chester, PA.: Schiffer Publishing Ltd., 1988.

Time-Life Books. *Encyclopedia of Collectibles: Dogs to Fishing Tackle*. Alexandria, VA.: Time-Life Books, 1978.

Timpson, Anne B. "The Anna Kempe House: German Neo-Classicism in Miniature." *International Dolls House News*. April/May 1996, pp. 38-41.

_____. "The Christian Hacker Firm." *International Dolls House News*, Winter 1993, pp.36-39.

_____. "Rococo Revival by Rock and Graner." *International Dolls House News*, December 1995/January 1996, pp. 39-41.

Towner, Margaret. *Dollhouse Furniture*. Philadelphia, PA.: Running Press, 1993.

Toy and Miniature Museum of Kansas City, Missouri. Kansas City, MO: The Museum, 1992.

The Universal Toy Catalog of 1924/1926. (Der Universal Speilwaren Katalog).Reprint Edition. London: New Cavendish Books, 1985.

Vaterlein, Christian. "The Marklin Story." Internet: http://www.marklin.com. n.p., n.d.

Whitton, Blair, ed. *Bliss Toys and Dollhouses*. New York: Dover Publications, Inc., 1979.

Whitton, Blair. *The Knopf Collectors' Guides to American Antiques: Toys*. New York: Alfred A. Knopf, 1984.

Whitton, Blair. *Paper Toys of the World*. Cumberland, MD: Hobby House Press, Inc.

Wisconsin Toy Company. Catalog circa mid-1930s. Milwaukee, WI: Wisconsin Toy Co. n.d.

INDEX

A

Arcade, 186-188
Airfix, 71
Althof, Bergmann and Co., 185
Amersham Works, Ltd., 50-52
Amloid Corp., 236
Andrews, O.B., Co., 204-205
Anglund, Joan Walsh, 212
Ardee Plastics, 221
Armstrong Cork Co., 214
Arrow Handicrafts Corp., 150
Atlantic Container Corp., 211
Auburn, 221-222

B

B. & S. Mfg. and Distributing Co., 73
Barbara Jean, 163
Barrett, A., and Sons, 57, 60
Barton, A., and Co., 57-66, 259
Bend-a-Family dolls, 265
Bex, 72
Biedermeier, 106-107, 109-110
Bisque dollhouse dolls, 251-257
Bliss, R., Mfg. Co., 6-23
Block House, 129-132
Blue Roofs, 76-83
Blue Box, 222-223
Box back dollhouses, 48-50
British Xylonite Co., 72
Brown, George W., and Co., 188
Built-Rite, 201-203

C

Caco, 260-261
Cairo, F., 34
Carol Lee Doll House Dolls, 265
Caroline's Home, 57-58, 61-66
Casablanca Products, Inc., 223
Cass, N.D., Co., 145
Charbens & Co., 72
China dollhouse dolls, 257
Cloverbloom Butter, 212-213
Cohn, Grete, 57, 259
Concord, 210
Converse, Morton E., 27, 145, 146-147
Cooke, Adrian, 189-190
Crescent Toy Co., 72

D

D.C. M.T., 72
Deauville, 83
Debbie's Dream House, 224
Deluxe Reading Corp., 224
Dennison's Crepe & Tissue Paper Doll House, 198, 200
Dinky Toys, 74
Dol-Toi Products, 67-71
Dolly Dear, 241-244, 249
Donna Lee, 164
Dowst Bros., 196-197
Dunham's Cocoanut, 29-30
Durrel Co., 208

E

Eagle Toy Co., 224-226
ECA Toys, 16454
Elgin, Eric, 37, 45, 46
Ermey, 263-264
Erna Meyer, 263-264
Evans & Cartwright, 72-73

F

Fairy Furniture, 189-190
Fairylite, 72-73
Farmer, Mark, 171, 265
Flagg Doll Co., 262
Frances, Mary, 172, 181
French metal furniture, 191, 197
Futureland, 236

G

Gable House and Carton Co., 208
GeeBee, 54
Gerlach, F.W., 122
Golden Oak, 107, 110-112
Goldilocks, 140-144
Gottschalk, Moritz, 76-99, 105
Graham Bros., 72-73
Grand Rapids, 181-182
Grandmother Stover's, 245-248, 262
Grecon, 57, 259
Grey Iron Casting Co., 194
Grimm & Leeds, 35
Gutter Houses, 23-27

H

Hacker, Christian, 100-103
Hall's Lifetime Toys, 124-129
Happi-Time, 210-211
Happy Hour, 166-168
Harco, 165
Hobbies, 52-54
House That Jack Built, 208-209
Hubley, 194

I

Ideal Home, 162
Ideal Novelty and Toy Co., 217, 226-227
Instant Play House, 212
Irene Miniatures Co., 227-229

J

Jacqueline, 73
Japan, 165-166
Jaydon, 230
Jayline Mfg. Co., 162
Jaymar Specialty Co., 166-168
Jenny's Home, 38, 222
Joy-toys Housekeeping Set, 199, 215-216

K

Kage Co., 168-170
Keggs, 163
Kenton Hardware Co., 194
Kestner, Johann Daniel, Jr., 106, 108
Keystone Mfg., Co., 148-150
Kiddie Brush and Toy, Co., 150
Kirkland, R.T., 241
Kleeman, O. & M., 73-74
Kleeware, 73-74
Kool Aid, 214

L

Lee Mfg. Co., 249
Lehman, E. & Co., 74-75
Lerch, A., 237
Lincoln, 170-171
Linemar, 195-196
Lines Brothers, 36-47
Lines, G. & J., 36-39
Little Hostess, 230-231
Little Orphan Annie Dollhouse, 150
Lowe, Samuel, 198, 205-207
Lundby, 58, 151
Lynnfield, 129-132

M

Marklin, 121-122
Marx, Louis, 166-167, 194-195, 230-231, 249
Mary Frances Housekeeper, 199
Mattel Creations, 235-236
McLoughlin Brothers, 31-34, 198-199
Meccano, 74
Menasha Woodenware Corp., 135-136
Meritoy Corp., 232
Messer, Sonia, 129-134
Miniaform, 172-173
Miniature Toy Co., 244
MinToy, 244
Miss Sunbeam's Doll House, 213
Mosher Folding Dollhouse, 17-18
Mystery Houses, 137-139

N

Nancy Forbes, 174-176

O

Ohio Art Co., 232

P

Packman, B., 176-177
Painted bisque dollhouse dolls, 258
Pennine, 54, 56, 57
Penny Toys, 191
Petite Princess, 226
Pia, Peter, 192
Pit-a-Pat, 74-75
Plasco, 233
Plastic Art Toy Corp., 233
Play-Town, 209

Q

Queen Anne, 176-177

R

Rallhouse, 233-234
Ralston Industries, 233-234
Rapaport Bros., 174-176
Realy Truly, 146-147
Red Lacquer, 107, 112
Red Line or Red-stained, 107, 113-114, 238
Red Roofs, 76-78, 84-94
Reed, Whitney S., 27-29
Reliable Plastics Co., 217-220

Renwal Mfg. Co., 227, 234
Rich Toy Mfg. Co., 151-153
Rock & Graner, 120-121

S

S&R, 191
Saalfield Publishing Co., 199, 201
Schmidt Lithograph Co., 199-201
Schneegas, Gebruder und Sohne, 106-107, 109-112
Schoenhut, A., 78, 93-94, 154-156
Schwarz, F.A.O., 137-140
Schweitzer, Babette, 122-123, 237
Sears Happi-Time, 210-211
Shackman, B. & Co., 178, 237, 261, 265
Shirley Temple Playhouse, 199- 201
Shotwell's, 214
Silber & Fleming type, 48-50
Simon & Halbig, 253-255
Simon et Rivollet, 191, 196-197
Star Novelty Works, 178-179
Stevens, J. & E., 192-194
Stoll & Edward Co., 208-209
Stover, John, 245-248
Strombeck Mfg. Co., 157
Strombeck-Becker Mfg., 156-157
Strombecker, 154, 156-157, 166
Strong, Ethel R., 262
Sunbeam, 213
Sutherland Paper Co., 209
Susy Goose, 150

T

Taylor & Barrett, 57, 60
Taylor, F.G. and Sons, 57, 60
Tekwood, 162
Thistle, 37, 43

Tinker Toys, 179-180
Tobacco rugs, 250
Tomac Toys, 260
Toncoss Miniatures, 180-181
Tootsietoy, 196-197.
Toy Tinkers, Inc., 179-180
Toyville Furniture, 216
Tri-ang, 36-47
Triangtois, 36-47
Trixy, 208
Tudor Toy Co., 54-56
Twink-L-Toy, 235
Twinky dolls, 262
Tyke Toys, 135-136
Tynietoy, 157-161, 237

V

Victory Toy Co., 172

W

Wagner, D. H. & Sohn, 103-105, 107
Waltershausen, 106-107, 109-112
Wanner, 181-182
Warren Paper Products, 201-203
Williams, A.C., 195
Winthrop-Atkins Co., 212
Wisconsin Toy Co., 140-144
Wonderland Doll's House, 199-201
Woodburn Mfg. Co., 164
Worth While Mfg. Co., 216
Wright, J.L., 170-171

Y

Young Decorator, 226